Textbook on
Rural Sociology and Educational Psychology

The Author

Dr. R. Velusamy is presently working as Associate Professor (Agricultural Extension) in the Department of Social Sciences, Horticultural College and Research Institute, Tamil Nadu Agricultural University, Periyakulam, Theni District, Tamil Nadu. During twelve years of service, he has significantly contributed in teaching, research and extension work and served in various positions in Tamil Nadu Agricultural University *viz.*, Subject Matter Specialist at Krishi Vigyan Kendra, Vriddhachalam, Cuddalore and as an Assistant Professor in the Department of Agricultural Extension and Rural Sociology, Agricultural College and Research Institute, Madurai. He has published two books and several research papers in journals of repute.

Textbook on
Rural Sociology and Educational Psychology

Dr. R. Velusamy

Associate Professor (Agricultural Extension)
Department of Social Sciences,
Horticultural College and Research Institute,
Tamil Nadu Agricultural University,
Periyakulam, Theni District, Tamil Nadu

2018

Daya Publishing House®

A Division of

Astral International Pvt. Ltd.

New Delhi – 110 002

Published by   : **Daya Publishing House®**
                           A Division of
                           Astral International Pvt. Ltd.
                           – ISO 9001:2015 Certified Company –
                           4736/23, Ansari Road, Darya Ganj
                           New Delhi-110 002
                           Ph. 011-43549197, 23278134
                           E-mail: info@astralint.com
                           Website: www.astralint.com

TAMIL NADU AGRICULTURAL UNIVERSITY
Horticultural College and Research Institute
Periyakulam – 625 604, Tamil Nadu, India

(Silver Jubilee Year – 2015)

Dr. V. SWAMINATHAN, Ph.D.,
Dean (Horticulture)

Foreword

Horticulture plays a vital role in the Indian economy. Horticulture crops occupy significant contribution in the total agricultural production in our country. Area under horticultural crops is increasing every year. Modem technology like namely, micro-irrigation systems, high density planting, hydroponics, Acropomics is suitable to horticultural crops. The new innovations in horticulture crops has to reach the farmers field for adoption in short period of time. The techniques and methods of technology transfer are very important to those who are involved in technology transfer process. Extension plays a vital role in transferring the new technology from lab to land. Sociological and psychological characteristics of farmers are important factors in adoption of new technology. This book deals with sociological and psychological characteristics of people, methods of data collection, importance of rural leadership and way of utilizing the local leader in technology transfer programme. In this way, this book will be very useful for the students of Agriculture, Horticulture and allied enterprise.

Dr. V. Swaminathan

Phone: **Off: 91-4546-231726**   Fax: **91-4546-231726**   Email:
deanhortpkm@tnau.ac.in   Website: **www.tnau.ac.in**

Preface

The innovations in agricultural and allied enterprises must reach the farmers for adoption to achieve the rural development. Extension workers play a vital role in transferring the technology from the research institutions to farmers field. Extension work is a change agent and they are trying to enrich the knowledge level of farmers, changing their attitude and increasing their skill. The rural people are having different settlement pattern, social status, cultural pattern, values etc,. The sociological and cultural aspects differ from individual to individual, place to place and region to region. The basic principles of extension is to educate the farmer in the aspects of how to think and not what to think. Extension worker should know individual farmers sociological aspects to plan effective technology transfer programme. Psychological components of farmers play a vital role in technology adoption. Studying the farmer's attitudes and planning the programme based on their attitude will leads to cent per cent adoption. The first eleven chapters deals the sociological aspects viz., Society, Social groups, Culture, Structure of rural society, Social stratification, Social value, Social control, Social change, Leader and leadership. The next six chapters deals with psychological aspects like Psychology, Educational psychology, Social psychology, basic principles of human behaviour, Intelligence, Personality, Motivation, Attitudes and teaching learning process. The last two chapters deal with data collection methods and training of leaders. The author expresses his heartfelt thanks to the several sources from which information has been drawn for writing this book.

Dr. R. Velusamy

Contents

Chapter 1

Sociology and Rural Sociology

Science is a body of organized and verified knowledge secured through scientific methods. Science has been classified into three types based on the nature of the subject matter; they are:

Physical Science

This branch of science deals with inorganic matter *i.e.*, the matter and energy having no reference to life. Examples are Physics, Chemistry, Geology, Astronomy, *etc.*

Biological Science

This branch of science studies organic matter *i.e.*, matter and energy having reference to life. Examples are Botany, Zoology, Entomology, Pathology, etc

Social Science

This branch of science deals with social phenomena/social life. Examples are Economics, Political Science, Psychology, History, Ethics, Anthropology, Sociology, *etc.*

Sociology

The term sociology was coined by *Auguste Comte* **(1789-1875)** who is often referred as the **father of sociology** who named it from **two** words, of which one is **Latin** word *socius* meaning **companion** and the other is **Greek** word *logos'* meaning **speech or reasoning**.

Sociology

Socius (Latin word)   Logos(Greek word)

Companion   Speech or reasoning

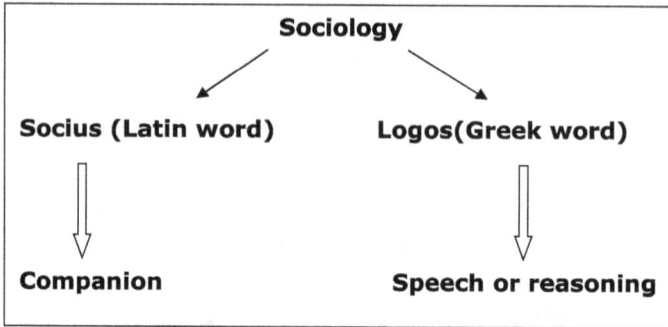

The etymological (based on the origin of the word) meaning of sociology is thus **'the science of society'**.

Definitions

Literally sociology means the study of the processes of companionship and may be defined as the study of the bases of social membership. No definition could entirely be satisfactory because of the diversity of perspectives which is characteristic of the modern discipline. However the definition for sociology according to various authors are given below:

☆ **Chitamber**: Sociology is the study of human beings in their group relations. As such it studies the interaction within and between groups of people.

☆ **M. Ginsberg**: Sociology is the study of human interactions and interrelations, their conditions and consequences'.

☆ **MacIver**: 'Sociology is the study of social relationships, which are referred to as 'web of society'.

☆ **Gillin and Gillin**: 'Sociology is the study of interactions arising from the association of human beings.'

☆ **Young and Mack**: 'Sociology is the scientific study of the structure of social life'.

☆ **Max Weber**: Sociology is the science which attempts the interpretive understanding of social man.

☆ **Rogers**: Sociology is defined as a study of ways in which social experiences function in developing, maturing, refreshing human beings through interpersonal stimulation.

☆ **J.B.Gitter**: "Sociology as the study of forms and processes of human togetherness". Sociology studies basic principle of human associations.

☆ **Fiddings**: "Sociology is the science of associations of minds; sociology tries to explain the origin growth and structure of society."

Although the definitions of Sociology vary in their focus, the common idea underlying them is that sociology is concerned with human relationships. Its subject matter is society rather than the individual, though the individual cannot be left out.

Since human life became complex, there was need for an in depth study of each aspect of human life. Thus, sociology has been further divided into different applied branches, namely; Rural Sociology, Urban Sociology, Political Sociology, Educational Sociology, *etc.*

Sociology adopts holistic approach in the study of human society *i.e.,* it studies all aspects of human life to generate a comprehensive knowledge in order to overcome problems of human life and society. This is why sociology is said to be a "social science par excellence"

Rural Sociology

Rural sociology is a **branch** of sociology. It is made up of two terms **rural** and **sociology** that is **science** of rural society. It is the study of the sociology of life in the rural environment, which systematically studies the rural communities to discover their conditions and tendencies and formulate the principles of progress as the term implies.

An extension worker is a **change** agent. Transfer or communication of innovations is the main job of these changes agents. But for introducing improved farm practices, an understanding of the farmer, his social and cultural environment within which he operates, his home, his village and the local region is necessary.

Rural sociology provides such knowledge and makes possible the planning of a strategic approach for the desired changes. It allows constant analysis of the rural situation and within reasonable limits prediction of possible results.

Definitions

'Rural sociology is the study of human relationships in rural environment.'

– Bertrand.

'Rural sociology is the scientific study of rural people in group relationships.'

– E.M. Rogers.

'Rural sociology is the scientific study of rural social relationships.'

– Lynn Smith.

Rural sociology is the study of sociological life in rural setting to discover their conditions and tendencies and to formulate principles of progress.

– A.R.Desai

Rural sociology is the body of facts and principles of the systematized knowledge, which has developed the application of scientific method in the study of human relationships in rural environment and people, engaged directly or indirectly in agriculture occupation.

– Smith

Nature of Rural Sociology

There is a controversy whether the sociology can be regards as science with his own subject matter. In the collection of facts for any knowledge, when we apply a scientific method it is called as a science. Science goes with the method and not with the subject matter. Scientific method consists of systematic observation, classification and interpretation of data. It is believed that rural sociology employs the scientific method. The nature of rural sociology as a science can determine on the basis of following facts:

1. Use of Scientific Method

It is uniform fact that rural sociology employs the scientific method. Almost all the methods of scientific study viz, observations, interview schedule, questionnaire method, case study, statistical methods *etc.* are employed in the study of rural sociology. In the absence of scientific approach the village problems can not be studied. On such studies we formulate generalized principles and laws on which we forecast future trends.

2. Factual Study

It studies the social events, social relationship and process in a factual manner. It also studies and analyzes the facts and the underlying general principles and theories.

3. Discovery of Cause and Effect

Rural sociology formulates its theories and laws on the basis of cause and effect relationship.

4. Universal Laws

The laws formulated by rural sociology are universal in nature. Because under normal or similar condition they prove to be correct and produce the same result.

5. Predictions

Since the laws formulated by rural sociology by rural sociology are based on cause and effect relationship. It is possible to predict the result.

On the basis of conditions enumerated above, it is said that rural sociology is by nature a science. However, there are some factors which limits the scientific sociology is by nature a scientific nature of the subject.

Limitations of Rural Sociology as a Science

1. **Lack of Objectivity**: It is not possible to have objectivity in the study of rural sociology as in case of natural science. While in the study of rural sociology and its problems the investigator continues to remain as a part of the society he is studying. He has own ideas and are influenced by the subject matter.

2. **Lack of Laboratory**: Rural sociology is not studied in laboratories as a natural science. Because of this it is not possible to verify and test the theory and principles of rural sociology.

3. **Lack of Measurement**: There is no definite and standard measurement for measuring units of rural sociology.

4. **Lack of Exactness**: It is lack of objectivity and different to follow its laws and principles universally. They are not acceptable at every point.

5. **Lack of Prediction**: Because of lacks objectivity and exactness the principles formulated by the rural sociology are not always correct. As such predictions are not possible.

6. **It is not possible to draw a line between the rural and urban areas**. There is no sharp demarcation to tell where rural area ends and urban area begins.

7. **The science of rural sociology is not fully developed.**

Scope of Rural Sociology

India is an agricultural country and village is a basic and important unit of the society. After independence the process of rural reconstruction was started and importance of the rural sociology was recognized. The need of development of the villages and speed of education in the villages were understood and to achieve this community development programme was started. It can be achieved when the planners and administrator correct knowledge of the rural life.

1. It is very essential to develop village because India's development depends upon the progress of the villages.

2. India is agriculture country and poverty can be removed through improvement in agriculture.

3. Solutions of rural problems can bring the change in the rural society

4. The country and its society can be reconstructed only through rural developments.

5. For successful implementation of democratic decentralization the village community is to be studied in detail.

6. Rural sociology can help to organize the disorganized Indian in detail.

7. The extension worker must know the rural culture, rural institutions, problems, resources *etc.* for successful transfer of technology for improvement of agriculture. It can be achieved through the study of rural sociology.

8. Through the technology and communication methods are known to the extension workers. The study of rural sociology helps the extension worker to transfer the technology.

9. For successful implementation of the community development programmes the knowledge of rural sociology is very essential.

Importance of the Rural Sociology in Extension Education

The importance of rural sociology can be evaluated properly when it realize the importance of rural society. Rural society presents a scientific picture of rural life. Villages are important because they are the springs to feed urban areas. The importance of rural sociology can be put under following heads.

Interrelationship between Rural Sociology and Extension

Sl.No.	Rural Sociology	Extension
1.	It is a **scientific study** of the laws of the structure and development of rural society	It is **informal** (actually non-formal) education for the rural people with a view to **develop** rural society on desirable lines
2.	It studies the **attitudes** and behavior of rural people	It seeks to **modify** or change for the better, the attitudes and behavior of village people
3.	It studies the **needs** and interests of rural society	It helps rural people to **discover** their needs and problems and **builds educational programs** based on these needs and wants
4.	It analyses rural **social relationships,** or group organizations and **leadership** in rural areas, the **social processes** like cooperation, association, competition etc, among village people	It **fosters** (develops) and utilizes **village organizations** and leadership and **favorable** social processes, to achieve its objectives of rural development
5.	It studies **social situations** and assembles social facts or rural society	It makes **use** of such **social data** as a basis for building up its **extension programs** for rural areas
6.	It **investigates** the social, cultural, political, and religious **problems** of rural society	It also studies these problems with reference to their **impact on extension work** in villages

Man has an urge to know human relationship and this can be satisfied through rural sociology.

a) **Rural Population is Majority**: In almost all the Countries of the world majority or the world resides in villages in villages. It is truer that over 80 per cent population of India resides in villages.

b) **It Gives Complete Knowledge of Village Life**: Rural sociology gives us complete knowledge of village life. Village is the first unit of development in country. It is a centre of culture of any country.

c) **Rural Reformation**: Rural reformation is the primary aim of rural sociology. In this context it helps in following works.

i) **Organization**: Village unit which are dis-organized and can be organized through rural sociology. It improved in the co-ordination of various units and helps in bringing an improvement in economic, social and health conditions.

ii) **Economic Betterment**: Through detailed study of village problems and observation rural sociology gives stress on the importance of increasing the quantity and quality of production. This results in to raising the standard of living.

iii) **Provide Technology and Systematic Knowledge and reforms in Farm Production**: Main occupation of 80 per cent population of village is agriculture. In order to improving this main occupation of rural people. The earlier researches in rural sociology was made in agricultural college.

iv) **Solutions of Social Problems**: Rural sociology examines the social pathological problems and it suggests ways for the improving these problem.

v) **Education**: The improvement, the development of any community depends on its education. Rural sociology lays stress on education in rural problems

vi) **Planning for Development**: Rural sociology encourages the development of various plans for any rural development programme. The work must be carried out according to these plans for the progress in rural society.

d) **Rural Sociology Development Relationships of Village with Industry.**

e) **Rural Sociology is most important in Agricultural Countries**: About 90 per cent of world progress is based on agriculture. It is only in agricultural countries that people realize the importance of rural sociology. India is mainly agricultural country. For its all sided development the development of rural sociology is very important.

Chapter 2
Society

Society is defined as a group of people in more or less **permanent** association who are organized for their **collective** activities and who feel that they **belong** together

A society is a collection of individuals united by certain relations and modes of behaviour which mark them off together who do not enter into these relations or who differ from them in behaviour.

– Ginsberg

A society may be defined as a group of people who have lived together long enough to become organised and to consider themselves and be considered as a unit more of less distinct from other human units.

– John F. Cuber

Elements of Society

1. Society means likeness

Likeness is an essential pre-requisite and therefore society means likeness.

2. Society also implies difference

If people are all very much like or exactly alike the social relationship becomes very much limited. Family rests upon the biological difference of sexes. There are differences in aptitude, interest and capacity. Difference is also necessary for society.

3. Inter-dependence

In addition to likeness, inter-dependence is another essential element.

4. Cooperation

Without cooperation no society can exist. They cannot have a happy life, (*e.g.*) Family rests on cooperation.

Rural

It is an area where the people are engaged in the primary industrial in the sense they produce things directly for the first time in co-operation with nature.

Urban

It is an area where the people are engaged in the secondary industrial in the sense they produce things not directly for the first time in co-operation with nature.

Characteristics of Indian Rural Society

1. **Agriculture** is main economic activity of rural people. It is based predominantly on **Agriculture. Agriculture** is the main source of livelihood. The land is distributed between certain families. The distribution of land is between a big land owner and rest of the community, possession of which (land) has prestige value

2. **Caste** is **dominant** institution of village. It is peculiar type of grouping found in rural India. The village is governed to a very great extent by traditional caste occupations, carpenters, cobblers, smiths, washer men, agricultural laborers etc all belonging to separate castes, caste relations are important characteristics of rural life

3. The **religious** and **caste** composition of village largely determines its **character**. Different castes exist in village due to **social distance**. The habitation of each caste is separated from others.

4. Each village is **independent**. All villages have their own organizations, authority and sanctions. Every village has **Panchayat** which is village **self government**

5. Village settlements are governed by certain **traditions**. The layout of the village, construction of houses, the dress etc is allowed according to the prescribed patterns of the culture of the area. In different areas a certain degree of diversity (differences between villages in the above aspects of the village life) in village organizations is **peculiar**

6. The rural society is **self-sufficient**. The unit of production in rural society is the **family**, which tries to produce much of its required goods. **Economic production** is the basic activity of rural aggregates (rural groups)

7. As a territorial, social, economic and religious unit, the village is a **separate** and distinct entity

8. It is common to find out a sense of **attachment** towards own settlement site. In rural society people do not have widely diversified tasks in different parts of the community

9. Village is characterized by **isolation**

10. The chief characteristic of rural life is **homogeneity,** there are not many differences among people pertaining to income, status *etc.*

11. The other characteristics are less density of population, less social mobility, less education, simplicity, traditionalism, fatalism, believing superstitions *etc.*

12. Women do not have full equality with men in several aspects of life.

Differences between Rural and Urban

Rural people are different from those living in urban areas. These differences are mainly due to the environment and its consequent impact on the lives of the people

Sl.No.	Particulars	Rural Community	Urban Community
1.	General environment and orientation to nature	Closely associated with nature. Direct effect of natural elements like rains, drought, heat, etc., on their lives	Remote from nature. Predominance of man-made environment
2.	Occupation	Major occupation is farming.	Most of the jobs are Non-agricultural occupations
3.	Working conditions	Being **agriculture** work in open air	Work in closed environment. Greater isolation from nature. Poor fresh air
4.	Family	Work as a unit. More unity or integrity and more contacts between members	Work in different occupations and contact is less between members
5.	Size of the community	'Agriculturalism' and size of community are negatively correlated. Community is small in size. Land to man ratio is higher	Large. Less land per person. Urbanity and size of community are positively co-related.
6.	Density of population	Low density of population	High density of population
7.	Material possession	Less	Different types and more
8.	Homogeneity and heterogeneity	More homogeneous. Similarity in social and psychological characteristics in the population. Such as beliefs, language etc,	More heterogeneous. Wide variety of interests, occupations, languages etc.
9.	Social institutions	Most of the institutions are a natural outgrowth of rural social life. Less of enacted (approved or created) institutions	Numerous enacted institutions
10.	Social stratification and differentiation	Less among groups and low degree of differentiation. Gap between higher and lower classes is less	Different types of groups like professional, occupational etc, and high degree of differentiation. Gap between the higher and lower classes is more
11.	Hierarchy	Less in number *e.g.* lower, middle and upper classes	More in number *e.g.* upper-upper, upper-middle, upper-lower, middle upper and so on

Contd...

Contd...

Sl.No.	Particulars	Rural Community	Urban Community
12.	Social contacts and type	Less number, social interaction is narrow. Primary contacts are more predominant. Personal and relatively durable relations. Man is interacted as a human	Large number, social interaction is wider. Secondary contacts are predominant. Impersonal, casual and short-lived relations. Man is interacted as number and address
13.	Social mobility	Occupational and territorial mobility is less intensive. Normally the migration current carries more individuals from countryside to the cities	Occupational and territorial mobility is found more intensive. Urbanity and social mobility are positively correlated. Only in the period of social crises migration is from cities to countryside
14.	Social control	Informal control *i.e.* more related to the values and traditions of the society	Formal control *i.e.* legally
15.	Social change	Rural life is relatively static and stable	Urban social life is under constant social change
16.	Social solidarity (unity)	Strong sense of belonging and unity due to common objectives, similarities and personal relationships	Comparatively less sense of belonging and unity due to dissimilarities and impersonal kinds of relationships
17.	Standard of living	Low standard of living,	High standard of living
18.	Educational facilities	Less	More
19.	Economy	Subsistence	Cash
20.	Communication and transport	Less transport facilities, bad roads etc	Many transport facilities, better roads, communication etc
21.	Society	A simple, uni-group society	A complex, multi-group society
22.	Culture	Sacred	Secular (all religions are equal)
23.	Leadership Pattern	Choice of leadership more on the basis of known personal qualities of individual, due to greater face to face contacts and more intimate knowledge of individual.	Choices of leadership is comparatively less on the basis of known personal qualities of individual

Though there are differences, there are also common attitudes and behaviours that both rural and urban people share as a member of larger culture. Common elements between urban and rural life is:

1. Common language, literature, philosophy *etc.*
2. Common institutions like religion, education, family life business and political organizations *etc.*
3. The residents of both village and cities enjoy equal rights and privileges.
4. The family system remains the same in both the cases to some extent.

Chapter 3

Social Groups

In the widest sense the word 'group' is used to designate a collection of items. Man is born in a **social group** and his first association is with his mother. From his birth until he dies man associates with groups in some way or other. Groups influence his attitudes, thinking and behaviour throughout his life. They deeply influence the development of this personality and play a vital role in his socialisation.

The consideration of the following terms will give a clear concept of what we generally meant by a' social group.'

☆ **Category:** Means collection of items that have at least one common characteristic that distinguishes from other items which have other characteristics in common, (*e.g.*) individuals between 15 and 20 years of age, for instance, are referred to as an 'age group'.

☆ **Aggregation**: It is a collection of individuals in physical proximity of one another, (*e.g.*) cinema audience, spectators of a football game. There may be some interaction between the individuals in an aggregation but it is generally of a temporary nature and lacks definite pattern of organisation. Interaction will be normally lacking.

☆ **Potential group:** It is a group made up of number of people having some characteristics common but does not possess any recognizable structure. A potential group may become a real group, if it becomes organised and comes to have a union or organisation. Students form a potential group as long as they have no union but once they become organised, they form a social group.

☆ **Social group:** It is a collection of two or more individuals in which there are psychological interactions and reciprocal roles based upon durable contacts, shared norms, interests, distinctive pattern of collective behaviour and structural organization of leadership and followership

Definition

According to **Chitambar** a **social group** is a unit of two or more people in reciprocal (to and fro) interaction or communication with each other

MacIever defined **social group** as a collection of human beings who enter into distinctive social relationships with one another

According to **Ogburn and Nimkoff** "Whenever two or more individuals come together and influence one another, they may be said to constitute a social group."

"A social group is a collection of individuals, two or more, interacting on each other, who have common objects of attention and participate in similar activities."

– Eldredge and Merrill

Characteristics of Social Group

1. **Relationship:** Members of group are inter related to each other. Reciprocal relations form an essential feature of a group.
2. **Sense of unity:** the members of the group are united by a sense of unity and a feeling of sympathy.
3. **We – feeling:** the members of a group help each other and defend their interests collectively.
4. **Common interest:** the interests and ideals of the group are common. It is for the realisation of common interests that they meet together.
5. **Similar behaviour:** The members of a group behave in a similar way for the pursuit of common interest.
6. **Group norms:** Every group has its own rules or norms which the members are supposed to follow.
7. **Social group is dynamic not static:** It changes its form and expands its activities time to time. Sometimes the changes may be swift and sudden, while at other times it may occur so gradually that its members are unaware of it.

Classification of Groups

1. Based on mode of Organization and functioning: Formal and Informal group
2. Based on Nature of the relationship/Interaction: Primary and Secondary group
3. Based on Nature of Membership/structure of membership: Voluntary and involuntary groups
4. Based on Size of the Group: Small and large groups

5. Based on Territorial Limitations: Nature territorial group; Artificial territorial group and Non – territorial group

6. Based on Profession of Occupation: Educational group, political group, artesian group *etc.*

7. Based on Duration: Permanent group and temporary group

8. Based on Social Class: Horizontal group and Vertical group

9. Based on Personal Feeling of Belonging: In group and out group

10. Based on the Type and Quality of Relationship: Gemeinschaft Group and Gesellschaft Group

Based on the Mode of Organisation and Functioning

A well-organized group has well defined objectives. The functioning of such groups is governed by number of rules and regulations. According to this criterion a group can be classified into two types.

Formal and Informal Group

Formal Group	Informal Group
These are formally organized and have prescribed structure *i.e.* constitution by-laws etc. These groups have more rules and regulations to govern their functioning. The relationship of the members is governed by these rules. These groups are generally large in size and the members have many restrictions	These are not formally organized and lack prescribed structure. In those types of groups there are no much formalities, rules and regulations. The degree of organization is less in the sense that the members have not be undergo confinements and strict limitations for behaviour and actions. The members have many liberties and very less control as exists in Friendship
E.g. Labour union, village council, students union, College, Government Departments, Army *etc.*	*E.g.* family, friendship group, play group *etc.*

Based on Nature of the Relationship/Interaction

Primary Group	Secondary Group
Primary groups are those whose members have personal, intimate and face-to-face relationships.	Secondary groups are the one whose members have non-intimate, impersonal and indirect relationships.
Examples are family, village, tribe, small neighborhood, peer group, play group, etc.	Examples are political party, city, trade union, nation, etc

Difference between Primary Group and Secondary Group

Primary Group	Secondary Group
Small in size, often less than 20 to 30 persons	Large in size
Personal and intimate relationships among members are there	Impersonal and aloof(distant) relationships among members
Much face to face association is there between the members	Less face to face contact

Primary Group	Secondary Group
Permanency is there and members are together over a long period of time	Temporary in nature. Members spend relatively little time together
Members are well acquainted and have a strong sense of loyalty or 'we' feeling and a strong amount of group pressure is present	Members are not well acquainted and anonymity prevails
Informality is most common *i.e.* group does not have any name, officers etc	Formality prevails *i.e.* group often has a name, officers and a regular meeting place
Group decisions are more traditional and non rational	Group decisions are more rational and the emphasis is on efficiency
Primary groups are **relationship** directed	Secondary groups are **goal** oriented

Based on Nature of Membership/Structure of Membership

Here nature or type of membership is taken into consideration *e.g.* whether the membership is optional or voluntary or compulsory. This depends upon the mode of entrance to the members. The members limit themselves to their own interest.

Voluntary Group	Involuntary Group
In these type of groups, the membership is voluntary and members have no compulsion to participate in the activity of the group. The withdrawal from such groups is also on voluntary basis.	In these types of groups membership is compulsory and members have no choice. Social conversions and traditions rather than personal choice determine the relationship as observed in a family. Every one borne in a family has to function as a member of a family by compulsion
e.g. friendship group, play group *etc.*	*e.g.* family, neighbourhood, caste, community nation etc.

Based on Size of the Group

Small Group	Large Group
Member in this type is considerably less. Each member an identify each other and can establish close or direct relationship. In such groups feeling of co-operation and sympathy can be achieved individually. Size is limited *i.e.* Less than 30	This type of group is bigger in size. Number of members in a group is considerably large. *I.e.* 61 to 1000. In larger group relationship is not direct and involves in to association.
e.g. Family, play group *etc.*	*e.g.* Political group, labour union University, Army etc.

Based on Territorial Limitations

Here the limit territory in which the group is functioning is taken into consideration. These groups work in defined territorial limits and they have fixed boundary.

Natural Territorial Group	Artificial Territorial Group	Non- Territorial Groups
These are the groups were territorial limits have been fixed by nature. Those groups are formulated by their natural similarities and boundaries. There boundaries and limits are fixed by geographic and climatic situations.	Here the territory is fixed artificially by man on functional basic	Here the natural and artificial territorial limit do not play any part. This group has a common functioning in all most all parts of the world and they do not take into consideration any limitation of other types
e.g. Region.	*e.g.* Village, Taluka, District, State *etc.*	*e.g.* UNO, FAO, Red Cross, International Trade Unions *etc.*

Based on Profession of Occupation

These groups are formulated on the basis of the professional or the occupations of the members *e.g.* Religious, Educational Groups, Political Groups, Artisans *etc.*

Based on Duration : The stabilized and short duration group can be of two types

Permanent	Temporary
The members remain together for great length of time and aware of the objective of the group. There are formalities and defined roles to play. The members are tied together by potential ties and formulates. The groups as **a** whole has permanent existence	Temporary groups are the mere collections of physical bodies and congregate in casual way on the street or on the stations. The size of such group is indefinite and they are unorganized. They do not remain together for great length of time. The participants are all on one level because their attention is focused on one thing and interaction is uncontrolled.
e.g. Family, Government Department *etc.*	*e.g.* Crowd, Audience, and Mob.

Based on Social Class

Based on **social class** groups are divided in the horizontal and vertical groups:

Horizontal Group	Vertical Group
These groups often organize themselves of a personal from the same level of society. The persons are alike in the status or position in the class system of society. Farmers, blacksmith, carpenter would be the members of their respective occupations belong to a horizontal group.	Vertical groups are those groups that are composed of members from different social strata. Its membership cuts vertically across the horizontal groupings in society *e.g.* political parties. The persons of different classes (*i.e.* lower and upper) work together in close relationship to promote their parties interest.
e.g. caste	*e.g.* race, nation *etc.*

Based on Personal Feeling of Belonging

Based on **personal feelings** the groups have been divided in and out groups:

In Group	Out Group
In-groups are the one to which people feel that they belong. Such groups are characterized by the expressions "We belong, we believe, we fell, we act or my family my neighborhood, my club, my association. In-group attitude contra constrains some elements of sympathy and sense of attachment or obligation to the other members of our group.	Out-groups are the one to which people do not feel that they belong. Every group is conscious that other groups are those to which we do not belong or not with us. We are democrats. They are Communist, we are Hindus, we are Muslims, we are Brahmins, and they are Harijans. A person has no sense of loyalty, sympathy, co-operation while they have sense of indifference even antagonism to the members of out-groups
e.g. my family, my class, *etc.*	*e.g.* their family, their class, *etc.*

Based on the Type and Quality of Relationship

Gemeinschaft Group	Gesellschaft Group
Here is the society which most relationship are traditional or personal or after both. In which landlords had his tenants who were personally known to him and who had obligation for their welfare and to whom tenants fulfilled certain obligations. In such groups written documents or contracts were not present while traditional pattern existed and was accepted by society	It is a society in which neither personal attachment nor important or traditional rights obligations and duties. Relationships are based on bargaining and clearly defined agreement. This society flourishes in urban area and business organizations or associations of whole salers.
e.g. Zamindar system in India.	*e.g.* Labour union society

Reference group is the group which the individual refers for advises on different aspects. An individual may have different reference groups for different purposes. In rural society the individual belongs to a comparatively small number of groups (largely primary) and his behaviour is largely determined by them. Reference group like friendship group may influence a farmer to accept or reject the adoption of an improved farming practice.

In this group the individual feels **identified** with the group but he **may or may not** be the member of the group, the group influences individual. He shares the objectives of this group, which he accepts. The reference group provides the **standards** that guide behaviour even when the standards are contrary to earlier membership groups. To understand the behaviour of human beings we must know their reference groups.

Chapter 4

Culture

Culture is an integral aspect of human life, which influences the attitudes, actions and patterns of living of people in a society. Culture is the sole possession of human beings and upon which they could be distinguished from other lower animals.

The extension education brings about the changes in the behavior complex of the rural people. The behaviour is in turn influenced by the cultural factors, extension workers, therefore should have knowledge of the culture of the rural people.

In social science, culture refers to totality of what is learned by individuals as members of the society. Culture is a way of life, mode of thinking, acting and feeling. Culture refers to the distinct way of life of a group of people, a complete design of living.

Definitions

According to Maclver "Culture is the expression of our nature, in our modes or living and thinking, inter course, in our literature, in religion, in recreation and enjoyment".

"Culture is that complex whole which includes knowledge, belief, art, morals, laws, customs and any other capabilities acquired by man as a member of society."

– E.B. Tylor

Cuber defined "Culture is the continually changing patterns of learned behavior and products of learned (including attitudes, values, knowledge and material objects) which are shared by and transmitted among the member of society".

Types of Culture

There are two types of culture; they are:

1. **Material culture** refers to concrete objects created by human beings to satisfy their desires. These are external to human beings. Examples are tools, furniture, automobiles, buildings, utensils, *etc.*

2. **Non-material culture** refers to an abstract creation of man such as ideas, customs, beliefs, habits, morals, laws, knowledge, *etc.* These are internal to human beings and are acquired through process of learning.

Characteristics of Culture

1. **Culture is an acquired quality**: Culture is learned or acquired after birth and through life consciously or unconsciously through agents influencing directly or indirectly on individuals. It is not innate.

2. **Culture is social heritage and not individual heritage**: It is inclusive of the expectations of the members of the groups. It is a social product. Culture is linked with past. The past endures because it lives in culture. It is passed from one generation to another through tradition and customs.

3. **Culture is idealistic**: Culture embodies the ideas and norms of the group. It is a sum total of the ideal patterns and norms of behaviour of a group.

4. **Culture fulfils some needs**: Culture fulfils those ethical and social needs of the groups which are ends in themselves.

5. **Culture is an integral system**: Culture possesses order and system. Its various parts are integrated with each other and any new element which is introduced is also integrated.

6. **Language is the chief vehicle of culture**: Man lives not only in the present but also in the past and future. He is enabled to do because he possesses language which transmits to him what was learnt in the past and enables him to transmit the accumulated wisdom.

7. **Culture is adoptive**: It must adjust itself to external forces of various kinds.

8. **Only human beings** posses the culture, other animals do not posses it. Man has created the culture during the process of controlling himself, others and nature.

9. All the societies in the world have culture but each society has a different **culture from one another**. Indian culture - African culture - Western culture.

Cultural Concepts

There are different aspects in the culture of a society, which are addressed by different concepts. The understanding of these concepts helps to understand the various dimension of culture.

1. **Cultural Traits** are the individual elements or smallest units of a culture. These units put together constitute culture. Thus, shaking hands, touching

the feet of elders, tipping hats, saluting the national flag, wearing white 'saris' at mourning, taking vegetarian diets, walking barefooted, sprinkling water on the idols are cultural traits. Thus, traits are the elemental units of a culture. It is these traits which distinguish one culture from another. A trait found in one culture may have no significance in other culture.

2. **Culture Complex** is a group or cluster of related culture traits. According to Hoebel "cultural complexes are nothing but larger clusters of traits organized about some nuclear point of reference." Cultural traits do not usually appear singly or independently. They are customarily associated with other related traits to form culture complex.

3. **Cultural Diffusion** is the process by which cultural traits spread from one group or society to another. This may take place either due to the physical proximity of people of different societies or through mass media of communication like T.V. *etc.*

4. **Cultural Relativity** (Cultural Relativism) is judging a culture on its own terms and not in comparison to another culture. In other words, judging the cultures in their own terms rather than by the standards of other culture is cultural relativism. Cultures should be judged only in the context in which they occur.

5. **Cultural Lag** is a situation in which some parts of culture (usually material culture) change at a faster rate than other parts (usually non-material culture). This concept has been given by Ogburn.

6. **Cultural Pluralism** is the living together of people despite cultural differences with sympathetic consideration to each other.

7. **Cultural Universals** are those cultural traits that apply to all the members of a society. Examples are incest taboo, respecting the national flag, loyalty and patriotism to the nation, respecting elders/women, *etc.*

8. **Cultural Alternatives** refer to the cultural traits that offer socially acceptable choices. Examples are different modes of dress, worshiping, occupation, customs, *etc.*

9. **Cultural Change** is the process of alterations in different spheres of culture of a society. This takes place due to two sets of factors, namely; (1) inventions and discoveries, (2) cultural diffusion and borrowings.

10. **Ethnocentrism:** Ethnocentrism is the tendency of a society to consider its own culture as best and others as inferiors.

Functions of Culture

Culture is important for individual and group. So one has to consider the functions of culture under two heads.

a. For individual and
b. For groups

a. For Individual

1. Culture makes man a human being, regulates his conduct and prepares him for group life. It provides him a complete design of living. It teaches him what type of food he should take and in what manner; how he should behave with his fellows; how he should speak with and influence the people and how he should co-operate or compete with others. In short, the qualities required to live in a social life are acquired by man from his culture.

2. Culture provides solutions for complicated situations. Culture provides man with a set of behaviour even for complicated situations. Culture thoroughly influences him so that he does not require any external force to keep himself in conformity with social requirements. His actions become automatic (*e.g.*) Forming a queue when there is a rush.

3. Culture provides traditional integrations to certain situations. (*e.g.*) if a cat crosses his way he post-pones the journey. These traditional interpretations differ from culture to culture.

b. For Group

1. **Culture keeps social relationships intact:** Culture is important not only for man but also for the group. Had there been no culture there would have been no group life. By regulating the behaviour of people and satisfying their primary drives pertaining to hunger, shelter, and sex it has been able to maintain group life. Infact life would have been poor, nasty and short if there had been no cultural regulations. It is culture which keeps all social relations intact.

2. **Culture broadens the vision of the individual:** Culture has given a new vision to the individual by providing him a set of rules for the cooperation of the individuals. It provides him the concepts of family, state, nation and class and makes possible the coordination and division of labour.

3. **Culture creates new needs:** Culture creates new needs and new drives for example, thirst for knowledge and arranges for their satisfaction. It satisfies the aesthetic, moral and religious interest of the members of the group. In this way groups owe much to culture.

Acculturation

The process and implications of cultural change. Individual or group-level change that occurs as a result of first-hand contact with another culture

Marginal Man

Marginal man or marginal man theory is a sociological concept first developed by sociologists Robert Ezra Park (1864-1944) and Everett Stonequist (1901-1979)) to explain how an individual suspended between two cultural realities may struggle to establish his or her identity

Ethos

Ethos is a Greek word meaning "character" that is used to describe the guiding beliefs or ideals that characterize a community, nation, or ideology.

Chapter 5

Structure of Rural Society

Social structure is a pattern or arrangement of elements of a society in an organized and collec-tive way. The interactions and behaviour of the members of a society are stable and pat-terned. These stable patterns of interaction are called 'social structures'.

Social structure is the framework of society that sets limits and establishes standards for our behaviour. It is, thus, defined simply as any recurring pattern of social behaviour. A social structure includes or is made-up of elements of society, such as institutions, statuses, roles, groups and social classes.

Pattern of Rural Settlement

Pattern of settlement has been defined as the relationship between one house or building and another. The rural settlements have different shapes and sizes. The site of the village, and the surrounding topography and terrain influence the shape and size of a village. In fact, the pattern of rural settlement is the result of a series of adjustments to the environment which have been going on for centuries. Moreover, socio-cultural factors such as caste structure of the people living in a village and the functional needs of the people also have a close bearing on its shape and size.

Type refers to a category of things having some common features whereas pattern refers to a regular form or order in which a series of things occur.

Types of Rural Settlements

1. Compact/clustered/nucleated settlement
2. Semi-compact/Semi-clustered/fragmented settlement

3. Hemleted settlement
4. Dispersed settlement.

1. Compact Settlements

As the name suggests, these settlements have closely built up area. Therefore in such settlements all the dwellings are concentrated in one central sites and these inhabited area is distinct and separated from the farms and pastures. Maximum settlements of our country comes under this category. They are spread over almost every part of the country.

Very often these settlements have a definite pattern due to closely built area and intervening street patterns. As many as 11 patterns are identified. We will discuss only Five major patterns. These patterns are: (i) Linear pattern (ii) Rectangular pattern (iii) Circular pattern (iv) Square pattern (v) Radial pattern

- (i) **Linear Pattern**: It is commonly found along main roads, railways, streams, *etc*. It may have a single row of houses arranged along the main artery. For example rural settlements found along the sea coast, river valley, mountain ranges *etc*.

- (ii) **Rectangular Pattern**: This is a very common type which develops around the rectangular shape of agricultural fields as it is common to find a system of land measurement based on square units. Village paths and cart tracks also confirm to the rectangular field patterns and run through the village in north-south and east-west directions.

- (iii) **Square Pattern**: This is basically a varient of rectangular type. Such a pattern is associated with villages lying at the crossing of cart tracks or roads and also related to features restricting the extension of the village outside a square space. These features may include an old boundary wall, thick orchards, a road or a pond.

- (iv) **Circular Pattern**: The outer walls of dwellings adjoin each other and present a continuous front so that when viewed from outside, the villages look like a walled and fortified enclosure pierced by a few openings. The round form was a natural outcome of maximum aggregation for the purpose of defence during the past.

- (v) **Radial Pattern**: In this type, a number of streets converge on one centre which may be a source of water (pond, well), a temple or mosque, a centre of commercial activity or simply an open space.

2. Semi-Compact Settlement

As the name suggests, the dwellings or houses are not well-knitted. Such settlements are characterized by a small but compact nuclears around which hamlets are dispersed. It covers more area than the compact settlements. These settlements are found both in plains and plateaus depending upon the environmental conditions prevailing in that area.

3. Hamleted Settlements

These type of settlements, are fragmented into several small units. The main settlement does not have much influence on the other units. Very often the original site is not easily distinguishable and these hamlets are often spread over the area with intervening fields. This segregation is often influenced by social and ethnic factors.

4. Dispersed Settlements

This is also known as isolated settlements. Here the settlement is characterized by units of small size which may consist of a single house to a small group of houses. It varies from two to seven huts. Therefore, in this type, hamlets are scattered over a vast area and does not have any specific pattern. Such type of settlements are found in tribal areas

Rural Social Institutions

Social institution is the structure and machinery through which a human society organizes, directs and executes the multifarious activities required to satisfy human needs. When man relates himself with others he creates what have been described as forms or structures in order to enable him meet his needs and function in other ways of life.

Definitions

1. "Institution is an established forms or conditions of procedure characteristic of group activity" – MacIver.
2. "Institution is the normative order of defining and governing the patterns of social action, deemed by the members of the group or society as morally and socially crucial to the existence of group or society" – Park and Burgess.

There are five major institutions in rural society.

1. Family
2. Educational institution
3. Political Government Institution
4. Religious
5. Economic (Occupation)

1. Family

Family is a universal primary social institution. It functions as a social, biological and economic unit and therefore it has a prime place in human society.

Definitions

1. "Family is a group defined by sex relationship sufficiently precise and enduring to provide for the procreation and upbringing of children"- MacIver.

2. "Family is a more or less durable association of husband and wife with or without children, or a man or woman alone with children"-Nimkoff.

Classification of Family

The families can be classified according to various criteria:

a) On the Basis of Lineage

1) Patrilineal Family: When property inheritance and reckoning descent along the male line (father).
2) Matrilineal Family: When it is along the female (mother) line.

b) On the Basis of Headship

1) Patriarchal Family: In this case father is head of the family.
2) Matriarchal Family: In this case mother is head of the family.

c) On the Basis of Transfer of Bride Groom

1) Patrilocal Family: There are the families where wife transfer to the husbands house after marriage.
2) Matrilocal Family: Where husband transfer to the wife's house after marriage. Matrilineal families are matrilocal families.

d) On the Basis of Number of Mates

1) Monogamous Family: In this case the husband marries only one wife.
2) Polygamous Family: In this case the husband can marry more than one wife.
3) Polyandrous Family: In this case the wife can marry more than one husband.

Nuclear or Conjugal or Individual Family

Such a family consists of married couple and their children, and is well separated from other relatives who may pay short visits if at all.

Consanguineal Family

Such a family consists often of grandparents, their sons, their sons 'wives and even their sons' grand children. Consanguineal literally means "of one blood". Eldest male member is the head of family.

2. Educational Institution

Educational Institutions are those which seek to socialize individuals in society. Every new generation must be prepare and trained to play a role in society. Education is necessary for social and economic development of man in society.

What is Education?

Education can be defined as production of desirable changes in the behavior of people or individual. The changes which are not desirable are not considered as a education.

Education is important for all. If village people are properly educated, they shall be able to know, what is expected of them and what way, they can be useful members of the society.

Religious Institution

Religion exists in every society in some form or other. Religion plays very important part in the life of the people on the whole in India. India is basically country of religious. Every aspect of life is governed by the religion. Rural society or the village people are more favorable inclined towards religion. Religion is very old institution. It is not clear as how and when it originated.

Definitions

"Religion is a set of beliefs regarding the relationship of a man to the supernatural power called God". Green says, "Religion is a system of belief and symbolic practices and objects governed by faith rather than by knowledge, which relates to man to unseen supernatural realm beyond the known and beyond the controllable."

Social Organizations

Social organizations are important aspect of human life and thus of society. Human beings are not only born, live in and work through social organizations but also satisfy most of their desires through them. Social groups that have been deliberately and consciously constructed in order to seek certain specific ends are called social organizations. In other words, social organizations are groups of people organized to pursue specific objectives.

Definitions

"Social organization is an articulation of different parts performing different functions, it is an active group devised for getting something done"

– Ogburn and Nimkoff

"Social organization is the patterned relations of individuals and groups. It is one of the sources of order in social life."

– Leonard Broom and Phillip Selznick

The form and structure of a social organization develops as a specialized activity, rules and regulations for operating, time and place of meetings are formulated, and the organization operates as a clearly defined entity having a specific objective with officers and membership. Examples of social organizations are school, political party, youth club, village panchayat, co-operatives, self-help group, trade union, *etc.*

Types: Social organizations are classified into four types based on different bases; they are:

1. Prescribed and Voluntary Social Organizations on the Basis of Political Structure

The prescribed organizations are primarily or wholly government established and controlled with membership being compulsory. Generally, they exist in totalitarian societies. Voluntary organizations arise as spontaneous expressions of interests of people in society with membership not being compulsory. These are generally found in non- totalitarian societies.

2. Recreational and Service Social Organizations on the Basis of Motives of Participation of People

The recreational social organizations clubs are the one in which people participate for personal pleasure and satisfaction. The service social organizations are formed to provide service for the welfare of members or the general public.

3. Open and Secret Social Organizations on the Basis of Organizational Operation

Open social organizations are the one whose programmes, goals and membership are known to the general public. Examples are political party, village panchayat, farmers' association, college, *etc.* Secret social organizations are those whose purposes, programmes, membership or activities are known only to the members. Examples are caste and religious organizations.

4. Inclusive, Restricted and Exclusive Social Organizations on the Basis of Admission to Membership

   a) Inclusive social organization is one in which membership is open to anyone interested in the purposes of the organization and meets its requirements. Examples are recreational clubs.

   b) Restricted social organization is one whose membership is open to the persons who possess predetermined qualifications that characterize the organization and the individual.Examples are associations of radio engineers, agronomists, doctors or textile manufactures, *etc.*

   c) Exclusive social organization is one in which the admission is given through election. Examples are rotary international, red crosssociety, *etc.,* where membership is at the discretion of the organization within its constitutional provisions.

Ecological Entities – Society, Community, Neighbourhood Elements, Relationship and differences

An entity is something that exists by itself, although it need not be of material existence. A person, partnership, organization or business that has a legal and separately identifiable existence.

1. Society
2. Community
3. Association

Society

A society, or a human society, is a group of people involved with each other through persistent relations, or a large social grouping sharing the same geographical or social territory, subject to the same political authority and dominant cultural expectations. Human societies are characterized by patterns of relationships (social relations) between individuals who share a distinctive culture and institutions; a given society may be described as the sum total of such relationships among its constituent members.

Having studied and understood the sociology as the science of the society, it is to investigate now what is society and its relationship with the individual.

Important aspect of society is not the structure; it is the system of relationship. The relations which do not have definite associations have been excluded from the definitions. Society exists only when the members know each other and possess common interest on subjects.

Society is defined as a group of people in more or less **permanent** association who are organized for their **collective** activities and who feel that they **belong** together

A society is a collection of individuals united by certain relations and modes of behaviour which mark them off together who do not enter into these relations or who differ from them in behaviour.

– Ginsberg

A society may be defined as a group of people who have lived together long enough to become organised and to consider themselves and be considered as a unit more of less distinct from other human units.

– John F. Cuber

Elements of Society

Society possesses a number of elements. Following are the important elements or characteristics of society.

1. Society means likeness

Likeness is an essential pre-requisite and therefore society means likeness.

2. Society also implies difference

If people are all very much like or exactly alike the social relationship becomes very much limited. Family rests upon the biological difference of sexes. There are differences in aptitude, interest and capacity. Difference is also necessary for society.

3. Inter-dependence

In addition to likeness, inter-dependence is another essential element.

4. Cooperation

Without cooperation no society can exist. They cannot have a happy life, (*e.g.*) Family rests on cooperation.

5. Organization

There is an important factor of society. It is some kind of organization. In other words every society has its own individual and unique organization. Society always requires an organization for its formation.

6. Social Relationship

Society consists of social relations, customs, laws, mores *etc.* These social relations are intangible and unseen. People only feel or realise these relations. Thus they do not have any concrete form and therefore society is abstract. In this way abstractness is a significant ingredient of society.

7. We-feeling

Society is based on we-feeling which means a feeling of belonging together. This we-feeling makes society identifiable and distinct people in comparison to other. The we-feeling which can distinguish societies from one another.

8. Social Group

A society is the social group. It encompasses all other social groups that exist among the people.

9. Society is Dynamic

Society is not static. It is dynamic. Change is ever present in society. Changeability is an inherent quality of human society. No society can even remain constant for any length of time. Society is like water in a stream or river that for ever flows.

It is-always in a flux. Old men die and new one are born. New associations and institutions and groups may come into being and old ones may dies a natural death. Changes may take place in every society slowly and gradually or suddenly and abruptly.

10. Social Control

Society has its own ways and means of controlling the behaviour of its members. Co-operation exists in society. But side by side competitions, conflicts, tensions, revolts and suppressions are also there. They appear and re-appear off and on. They are to be controlled. The behaviour or the activities of people are to be controlled. Society has various formal and informal means of social control. It means society has customs, traditions, conventions and folkways, mores, norms and so on. All are the informal means of social control. Society has also law, legislation, constitution, police, court, army and so on. All are the formal means of social control to regulate the behaviour of the members of the society.

11. Comprehensive Culture

Each society is distinct from the other. Every society is unique because it has own way of life, called culture. Culture refers to the social heritage of man.

It includes the whole range of our life. It includes our attitudes, judgments, morals, values, beliefs, ideas, ideologies and institutions. Culture is the expression of human nature in our ways of living and thinking, in behaving and acting as members of society.

Community

Person rarely exists alone. It is inevitable that people who over a long period reside in a particular locality should develop social likeness, have common social ideas, belonging *etc.*

A community is a local grouping within which people carry out a full round of life activities.

Although families or other groups can sometimes be relatively self-sufficient, most of them do not live in isolation. For many reasons, ranging from economic interdependence to shared cultural values, families and other groups normally join together to form communities.

The community, rather than the family, becomes the social setting for most everyday economic, political, religious, educational, recreational and similar activities. As communities become larger and more complex, other types of organisations are often established within the community to perform these various functions.

Thus, a community is a type of social organisation that is territorially located and provides the setting for dealing with most of the needs and problems of daily living. Communities vary widely in size and complexity.

Definition of Community

Community is a human population living within a limited geographic area and carrying on a common inter dependent life.

– Lund berg

Community is a social group with some degree of we feeling and living in a given area.

– Begardus

"Community is the smallest territorial group that can embrace all aspects of social life".

– Kingsley Davis

Ginsberg defines Community as "a group of social beings living a common life including all the infinite variety and complexity of relations which result from that common life which constitutes it".

According to Parsons, "A community is that collectively the members of which share a common territorial area as their base of operation for daily activities".

Elements of Community

Following are the elements on the basis of which it run he decided whether a particular group is a community

i) Group of people

Community is a group of people

ii) Locality

That group as people reside in a definite locality, a territorial area

iii) Community sentiment

It means a feeling of belonging together. It is a "we feeling" among the members, in modern societies it lacks. Even the neighbours are not known. Mere neighbourhood does not create a community, if community sentiment is lacking.

iv) Permanency

Community is not transitory like a crowd. It refers permanent life in a definite place.

v) Naturality

Communities are not created by an act but are natural

There can be larger communities like nations and smaller communities like neighbours. Here the larger community provides peace and protection and the smaller community provides friends and friendship

The other characteristics of a community as follows:

1. Mutuality
2. Common values and beliefs
3. Organized interaction
4. Strong group feeling
5. Cultural similarity

Difference between Society and Community

Society	Community
Society is a web of social relationships. It includes every relationship which is established among people	Group of people who live together in a particular locality and share basic conditions of a common life.
It refers all social structural relations direct or indirect, organised or unorganised, cooperative or antagonestic	To constitute a community the presence of sentiment among the members is necessary.
Society involves both likeness and difference. Both common and diverse interests are present in society.	But, likeness is more important than difference in community

Society	Community
It has no definite boundary	It refers a group of people living together in a particular locality
Community exist within society and possess its distinguishable structure	Small communities exist within greater communities like villages or nagars within a town.
Sense of "we feeling" is not essential in a society	Community sentiment is indispensable for a community. There can be no community in the absence of community sentiment
Society is wider; there can be more than one community in a society	Community smaller than society.
Society is abstract. It is a network of social relationships which cannot see or touched	On the other hand, community is concrete. It is a group of people living in a particular area. We can see this group and locate its existence.
On the other hand, common interests and common objectives are not necessary in society.	In a community, common interests and common objectives are necessary. People in a community live together for achievement of common interests and common objectives

Association

A group of people organised for a particular purpose or a limited number of purposes. Sometimes it is a group of social beings attached to an organisation with a view to secure a specific end or specific ends.

Definition of Association

Mac Iver defines an association as "an organisation deliberately formed for the collective pursuit of some interest or a set of interests, which its members share".

According to Ginsberg an association is "a group of social beings related to one another or have instituted in a common organisation with a view to secure a specific end or specific ends.

Elements of Association

    i) There must be a group of people

    ii) It must be an organised one based upon rules and regulations

    iii) They must have a common purpose

Thus family, church, trade union, music or any such clubs are associations. Associations may be formed on several basis.

Vocational (trade unions, teachers association recreational (Tennis clubs) or philanthropy (Charitable societies). They may be formed on the basis of duration (*i.e.*) temporary or permanent like flood relief association or trade unions.

Difference between Society and Association

    1. Society is older than association: Society exists since man appeared on the earth, while association arouse when man learnt to organise himself.

2. Aim of society is general: Society is for the well being of the individuals, whereas association formed for the particular purpose or purposes.

3. Society may be organised or unorganised but association must be organised.

4. Membership of society is compulsory as no man can live without it. But man can live without being a member of any association at all. Society will exit as long as man exists but association may be only transitory.

5. Society is marked by both cooperation and conflict whereas association is based on co-operation alone.

6. Society is a system of social relationship, whereas association is a group of people.

7. Society is natural, whereas association is artificial.

Difference between Association and Community

1. An association is partial, whereas community is a whole. Association is to achieve some specific purposes, which does not include the whole purpose of life. Whereas community includes whole circle of common life. It is not for specific purpose.

2. Community is not deliberately created. It has no beginning, no hour of birth. It is simply whole life, more comprehensive and more spontaneous than any association.

3. Associations exist within in community. Human beings belong to association by virtue of some specific interests that they possess (music club *etc.*). Though they born into communities they choose their associations.

It should be remembered that associations may become communities by serving plurality of ends, though that may never be reached.

Neighbourhood

Neighbourhoods have been described as limited geographic areas in which the individuals and families are known to each and carry on intimate associations together.

Neighbourhood is a smaller unit than the community; a community often is compared of several neighbourhoods.

A neighbourhood may be characterised by:

1. A locality group of people

2. Limited geographical area

3. Frequent participation in common activities such as visiting, inter-dining, borrowing and exchanging and other forms of mutual aid and

4. Presence of some service or supply agency, organisation or institution.

Neighbourhoods are hence more sociability entities based on personal relationships than are communities, which are based on social and economic requirements of residents and the satisfaction of them. This does not imply that conflicts do not arise in neighbourhoods and that relationships are always amicable. Conflict may be quite intense within neighbourhoods and within family circumstances.

Neighbourhoods usually have homogeneity. There exists a physical closeness of dwelling places and greater inter –personal contacts among those living in the same neighbourhood. Within individual neighbourhoods, the men and women may associate together in both work and leisure time. Thus, the neighbourhood is important because of the effects of the interpersonal relationships and social interaction on the decision making process, both in the families and the community.

Chapter 6
Social Stratification

Social stratification refers to the division of society into different strata or ranking of people or groups into socially superior and inferior positions. Strata imply existence of status differences characteristic of groups or society.

Definitions

'Social stratification is the process by which individuals and groups are ranked in a more or less enduring hierarchy of status.

— Ogburn and Nimkoff

'Social stratification is the vertical division of society into different social strata. Strata imply different status levels

— J.S. Rucek and R.L.Warren

The term status is the position held by people or groups in society in relation to others. Examples of status are engineer, doctor, mother, adult, boy, girl, student, leader, *etc.* There are two types of status, they are ascribed status and achieved status.

Ascribed status is one given to individuals or groups based on certain factors such as caste, religion, sex, age, race, *etc.*, upon which they have no control:Examples are:

☆ Caste status – Brahmin, Harijan, *etc.*

☆ Racial status – White and Negro, *etc.*

☆ Religious Status – Hindu, Muslim, Sikh, *etc.*

☆ Age based status – Infant, child, adult, aged, *etc.*

☆ Sex based status – Boy, Girl.

Achieved status is one obtained by people or groups by their efforts and personal competence. Examples are professor, doctor, engineer, etc

Bases for Social Stratification

The important bases upon which people or groups are ranked into different social status levels are:

1. Income
2. Wealth
3. Education
4. Occupation
5. Caste
6. Gender or Sex
7. Race

Types of Social Stratification

There are two types of social stratification, namely;

1. Open social stratification
2. Closed social stratification

Open Social Stratification

Open social stratification is one wherein there is an opportunity for people or groups to move upwards or downwards in their status based on their efforts and personal competence. This stratification prevails very much in industrially advanced societies. Example of this social stratification is social class system.

Social Class

Social class refers to a group of people having more or less same status such as higher or middle or lower status. The social classes generally found in society are upper, middle and lower classes, which are based on the factors like income, wealth, education and occupation. In Indian rural society, we find different classes based on landholdings possessed by people. These classes are large farmers, medium farmers, small and marginal farmers and agricultural laborers. These classes of rural society are known as agrarian classes.

Characteristics

Social class possesses certain important characteristics, *viz.*

a) Social class is a status group.
b) Social class is a culturally homogeneous group.
c) Social class involves more of class consciousness.
d) Social class maintains social distance.
e) Social class is an open system.

Closed Social Stratification

Closed system of social stratification is one wherein people or groups do not have adequate opportunities to move from one status to the other. Rather, they are required to remain in that status which is given to them on the factors beyond their control. Example of this stratification is Indian caste system.

Caste System

Caste is an example for closed system of social stratification in Indian society. The term caste owes its origin to Spanish word 'Casta' which means breed, race or a complex of hereditary qualities. The Portuguese applied this term to the classes of people of India known by the name of 'Jati'. The English word caste is an adjustment of the original term.

Definition

"Caste is a group of people who often (not always) have association with hereditary occupation, eat and marry among themselves and avoid (minimize) interaction with members of other out-groups."

– M.N. Srinivas

Characteristics

Six important characteristics of caste system.

1. Segmental division of Indian society.
2. Social and religious hierarchy
3. Endogamy
4. Restriction on feeding and social intercourse.
5. Lack of unrestricted choice of occupation.
6. Civil and religious disabilities.

Differences between Class and Caste Systems

1. Class is an open system where as caste is a closed system.
2. Class is secular in nature whereas caste is divine in nature.
3. Class is non-endogamous whereas caste is endogamous.
4. There are no restrictions on food habits, interaction, occupation in class system while there are rigid restrictions in caste system in respect of food habits, interaction and occupation.

Chapter 7

Social Value

Values are relative importance or preferences we give to any object, idea or content of experience *etc.* Value is defined as anything desired or chosen by someone.

The function of extension is to bring about desirable changes in the behaviour of people. The overt behaviour what we call as action is based on attitude. Unlike action attitude cannot be seen. They can only be inferred by way of tendency to act or react positively or negatively to some stimuli. These tendencies or attitude in turn are based upon individuals values.

Definition

Social values are the attitudes held by individuals, groups, or society as a whole, as to whether material or non – material objects are good, bad, desirable or undesirable

– Chitamber

Social values are relatively enduring (lasting or permanent) awareness plus emotion regarding an object, idea or person.

– Green

Social values are abstract and often unconscious assumptions of what is right and important.

– Young

Values and Norms

Norms are closely associated with values but are clearly differentiated from them.

The rules that govern action directed towards achieving values are called norms. Norms are the accepted and approved forms of behaviour that are based on and consistent with dominant social values in society. Thus values and norms go together.

$$\text{Opinion} \longrightarrow \text{Attitude} \longrightarrow \text{Norm/Social value}$$

A set of social values will always have an accompanying set of social norms or rules that uphold and support values

e.g. of value: Religious worship and respect to god usually is considered value *e.g.* of value system: Religion.

Examples of Norms

Observance of religious festivals and performance of rituals and worship and other relevant activities are important norms of society towards the value system of religion.

Major Values Prevailing in Rural Society or Social Values in Indian Rural Society

1. **Importance of ascribed (given by somebody) status**: Status of individual is decided by the group to which he belongs. There is an established order of hierarchy of castes in the Indian society

2. **Recognition of inequality**: Caste is still a guiding factor. There are inequalities based on the concept of higher and lower castes which are manifested (brought out) in many ways

3. **Patriarchal tendency**: Father is the head of family. Eldest male member of family has supreme power and tends to act autocratically

4. **Status of women**: There is a tendency towards giving greater respect and recognition to women, but they are supposed to be inferior to men. As far as their sphere of work is concerned it is mostly restricted to home management

5. **Greater male dominance**: Boys receive greater attention than girls. *E.g.* it is general attitude of parents that daughter(s) need not be highly educated

6. Adherence to well regulated sex relations

7. **Charity**: There is religious significance and approval for the giving of alms (something or money or food item given freely to poor). A person with a charitable disposition is respected

8. **Tendency of non-violence**: Killing of animals expect for the purpose of food is considered to be immoral

9. **Respect for old aged and elders**: There are fixed norms which guide the behaviour of individuals towards elders, superiors and old persons

10. **Religious attitude**: People in rural areas are religious. Performance of rituals and ceremonies are common in the traditional way

Types of Values

1. **Ultimate values**: Ultimate values are often referred as **dominant values**. These values express the general **views of society** towards matters such as the nature of the universe and man relation to it and to his fellowmen. These values are found most easily in **social institutions** such as **religion**, government or the family. *e.g.* The democratic proceedings expressed in the system of government (democracy). Ultimate values are **abstract** (not specific) and often **not** attainable

2. **Intermediate values**: These values are derived from **ultimate values** and are actually ultimate values that have been rephrased into more reasonable attainable categories. *e.g.* **Freedom of speech**, adult franchise (choice, religious freedom, free public education, non-discrimination, adequate housing *etc.*)

3. **Specific values**: The subdivisions of **intermediate** values are called specific values and are almost **unlimited** in number. Specific values must be in conformity with the total value system of which they form the **smallest unit**. *e.g.* To a farmer with intermediate value of adequate housing the related specific values can be a brick construction with a **flat slab roof**, wide verandah and large court and with provision to livestock housing. If public education is the intermediate value specific values can be the type of school, room and other facilities and content of courses or instructions *etc.*

Characteristics of Values

1. Values are constructs of society created through the interrelationships of its members. They are socially created rather than determined biologically or inherited.

2. Values are socially shared. While individuals in society may have individual values, the set of values that constitute the value system of society are shared and transmitted among members and accepted by them.

3. Values are learned. They are acquired not inherited. The process of learning and acquisition of social values commences from childhood in the family and through the process of socialisation.

4. Values are abstract attitudes and assumptions on which there is a social consensus about relative worth of objects in society.

5. Values are gratifying to people and have an important part in meeting social needs.

6. Values tend to be linked together harmoniously to form pattern; these patterns form the value system in society.

7. Value system vary from culture to culture in accordance with the relative worth attributed by each culture to its patterns of activity and its goals.

8. Values frequently represent alternatives and value systems consists of ranked alternative. Value therefore compete with one another, and behaviour is determined by the ranked position or priority level of the value.

9. Values may differ in their effects upon the individual and society as a whole. The values of a sub-group within society may be in conflict with those of society as a whole and work against its interests and welfare.

10. Because of their importance to individuals and society, values involve emotions, and people often sacrifice and even enter into conflict to uphold them.

11. Values exert strange influence on the development of individuals and society in at least two important ways. First by making it easy or difficult for rural people to accept new practices to form new type of organisations and operate in new ways. Second by influencing the scientific findings of the rural social scientist.

While values and norms are learned, many become internalised and form a part of the subconscious of the individual.

Chapter 8
Migration

Migration (human) is the movement of people from one place in the world to another for the purpose of taking up permanent or semipermanent residence, usually across a political boundary. An example of "semipermanent residence" would be the seasonal movements of migrant farm laborers. People can either choose to move ("voluntary migration") or be forced to move ("involuntary migration").

Types of Migration

☆ **Internal Migration:** Moving to a new home within a state, country, or continent.

☆ **External Migration:** Moving to a new home in a different state, country, or continent.

☆ **Emigration:** Leaving one country to move to another (*e.g.*, the Pilgrims emigrated from England).

☆ **Immigration:** Moving into a new country (*e.g.*, the Pilgrims immigrated to America).

☆ **Population Transfer:** When a government forces a large group of people out of a region, usually based on ethnicity or religion. This is also known as an involuntary or forced migration.

☆ **Impelled Migration** (also called "reluctant" or "imposed" migration): Individuals are not forced out of their country, but leave because of unfavorable situations such as warfare, political problems, or religious persecution.

☆ **Step Migration:** A series of shorter, less extreme migrations from a person's place of origin to final destination—such as moving from a farm, to a village, to a town, and finally to a city.

☆ **Chain Migration**: A series of migrations within a family or defined group of people. A chain migration often begins with one family member who sends money to bring other family members to the new location.

☆ Chain migration results in migration fields —the clustering of people from a specific region into certain neighborhoods or small towns.

☆ **Return Migration:** The voluntary movements of immigrants back to their place of origin. This is also known as circular migration

☆ **Seasonal Migration:**The process of moving for a period of time in response to labor or climate conditions (*e.g.*, farm workers following crop harvests or working in cities off-season; "snowbirds" moving to the southern and southwestern United States during winter).

Factors Influencing Migration

Economic Factors

Despite the relevance of non -economic factors most of the studies indicate that migration is primarily motivated by economic factors. In large number of developing countries, low agricultural income, agricultural unemployment and underemployment are considered basic factors pushing the migrants towards prosperous or dynamic areas with greater job opportunities.

Even the pressure of population resulting in a high man-land ratio has been widely recognized as one of the important reasons of poverty and rural out migration. Thus, almost all studies concur that most of the migrants (excluding forced and sequential migrants) have moved in search of better economic opportunities. This is an accepted fact in both internal as well as international migration.

The basic economic factors which motivate migration may be further classified as 'Push Factors' and 'Pull Factors'.

Push Factors

The **push factors** are those that compel a person, due to different reasons, to leave that place and go to some other place. For instance, low productivity, unemployment and underdevelopment, poor economic conditions, lack of opportunities for advancement, exhaustion of natural resources and natural calamities may compel people to leave their native place in search of better economic opportunities. In most developing countries, due to population explosion land -man ratio has declined resulting in significant increase in unemployment and underemployment. Introduction of capital intensive methods of production into the agricultural sector, and mechanization of certain processes reduce labour requirements in rural areas. The non-availability of alternative sources of income (non -agricultural activities) in rural areas is also important factor for migration. In addition to this, the existence of the joint family system and laws of inheritance, which do not permit the division of property, may also force many young men to move out to cities in search of jobs. Even sub division of property leads to migration, as the property become too small to support a family.

Pull Factors

The **Pull factors** refer to those factors which attract the migrants to an area, such as, opportunities for better employment, higher wages, facilities, better working conditions and amenities *etc.*

"Migration from the country side to the cities bears a close functional relation to the process of industrialization, technological advancement and other cultural changes which characterize the evolution of modern society in almost all parts of the world. Under the capitalistic model of development, there is a tendency for large proportion of investments to concentrate in the urban centers which encourage people to move to urban areas in the expectation of higher paid jobs. In recent years, the high rate of migration of people from India as well as from other developing countries to U.K., U.S.A., Canada and Middle East is due to the better employment opportunities, higher wages and the chances of attaining higher standard of living. Sometimes the people are also attracted to cities in search of better cultural and entertainment activities. Thus, pull factors operate not only in the rural -urban migration, but also in other types of domestic as well as international migration.

Demographic Factors

The differences in the rates of population increase between the different regions of a nation have been found to be a determinant in the internal migration. Fertility and the natural increase in population are generally higher in rural areas which drift the rural population towards the city.

Other important demographic factor in internal migration is marriage. The female migration is largely sequential to marriage, because it is a Hindu custom to take brides from another village.

According to National Sample Survey, more than 46 per cent migration to urban areas is caused by marriage. The custom of women returning to her parents to deliver her first child also accounts for significant internal migration.

Socio-Cultural Factors

Social and cultural factors also play an important role in migration. Sometimes family conflicts, the quest for independence also cause migration especially, of those in the younger generation. Improved communication facilities, such as, transportation, impact of the television, the cinema, the urban oriented education and resultant change in attitudes and values also promote migration.

Political Factors

Sometimes even political factors encourage or discourage migration from region to another. For instance, in India, the reservation of the jobs for 'sons of the soil policy' by the state governments will certainly discourage the migration from other states. Hence, the political background, attitudes and individual viewpoint of the people exercise a significant influence on the migration of the people.

Miscellaneous Factors

In addition, a number of other factors, such as the presence of relatives and friends in urban areas who mostly provide help, desire to receive education which is available only in urban areas are factors responsible for migration. Migration is considerably influenced by factors such as the closeness of cultural contracts, cultural diversity *etc.*

Chapter 9

Social Control

Social control is a universal phenomenon whereby social order and harmony are ensured in a group or society for its effective functioning by defining and regulating actions of members in accordance with its norms. Thus, social control refers to the process of influencing the members of a group or society to function in accordance with its norms and usages. The norms and usages refer to the rules of behaviors and practices accepted and followed by the group or society.

Definitions

"Social control refers to the process - planned or unplanned -whereby society or group influences its members to conform to its usages and values."

– Roucek

"Social control is sum of those methods by which society tries to influence human behaviour to maintain a given order."

– Mannheim

Need of Social Control

Social control is needed for a group or society to define and regulate behaviours of its members in order to ensure orderliness and harmony essential for its effective functioning.

E.A. Ross has given the following objectives of social control.

1. **To bring about social conformity**: Social conformity refers to the process of behaving in accordance with norms of a group or society. The opposite of social conformity is social deviance which refers to the process of behaving against norms of a group or society. Social conformity is brought about by

means of social sanctions by encouraging right behaviours by means of rewards and discouraging wrong behaviours by means of punishments. This promotes social equilibrium essential for effective functioning of a group or society.

2. **To promote social unity**: Social order and harmony become a dream without social control. Social control regulates human behaviour through established norms and consequently uniformity of behaviour develops leading to social unity. Unless unity prevails in a group or society, it becomes impossible for smooth social living.

3. **To promote social welfare**: Social control always aims at common welfare rather than welfare of an individual. An individual is influenced to act in accordance with the norms not for his benefit alone but also for the benefit of a group or society as a whole.

4. **To check cultural mal adjustment**: Culture provides certain patterns or ways of living for people to follow. Social control prevents people from going contrary to the ways of living given by the culture of a group or society. This process is known as checking cultural mal-adjustment. This is essential to maintain social harmony.

5. **To bring about continuity of a group**: The continuity of a group or society depends on the right behaviours of its members. Social control ensures the right behaviours by persuading the members of a group or society to adhere to its norms. Consequently, social order and harmony set in paving the way for continuation of a group or society. If the members of a group or society behave contrary to the group or societal norms disintegration of group or society sets in.

Forms of Social Control

There are two forms of social control; they are informal and formal social controls.

Informal Social Control

Informal social control refers to unwritten form of social control which emerges from collective life of people over ages and gets expressed through their collective reactions. These are non-deliberate but none the less effective. The important means of informal social control are customs, folk-ways, mores, taboos and rituals.

Customs

Customs are socially **prescribed** forms of behavior transmitted by tradition and enforced by social disapproval of its violation (not doing). Training the young, supporting the aged etc are some of the customs of society. Our acting, our dressing, our worship are controlled to a great extent by customs.

The **classification of customs** and their origin are as follows:

1. Folkways and usages
2. Mores and Taboos

3. Conventions
4. Rituals

Folkways

- ☆ Folkways are **expected** forms of behavior but are not rigidly enforced
- ☆ Folkways are the **customary** ways of behaving in society
- ☆ Folkways are **recognized** ways of behavior in a society
- ☆ The Folkways are socially **acceptable** ways of behavior. The customary norms of society that do not imply moral **sanction** (punishment). Folkways are otherwise called as **Usages** sometimes
- ☆ The folkways are the **right** ways to do things because they are the **expected** ways. People who do not conform may be subject to criticism but would not be penalized.

Examples of Folkways

- ☆ Good manners
- ☆ Entering home only after removal of shoes
- ☆ Lady touching the feet of her mother-in-law
- ☆ Greeting others with folded hands

Mores

Mores are the socially **acceptable** ways of behavior that do involve **moral standards** (regulations) and violation of **more** may result in severe **social action** or **sanction**, such as ostracism (exclusion of individual or family from the village or society). Religion provides foundation for **mores** of the society.

Examples of Mores

- ☆ Inter-dining of high-cast Hindus with out-caste Hindus
- ☆ Honesty is one of the recognized mores of the society
- ☆ Saluting the National Flag
- ☆ Standing during the playing of National Anthem
- ☆ Monogamy (having one wife or husband)

The term **more** is used for those things that are **ought to be done**. It is used for **positive** actions.

Taboos

Generally the term **'more'** is used for the **positive** action or things that ought to be done but the term **'taboo'** is used for the **negative** action and for the things that one **ought not to do**. Taboo means forbid. It refers to the prohibitions of the types of behavior because of some magical, supernatural (God) or religious sanction

Examples of Taboo

Total abstinence (self denial) of eating beef in a Hindu village (eating beef in Hindu religion) and eating pork in Muslim religion.

Rituals

Ritual may be defined as a pattern of behavior or ceremony, which has become the **customary** way of dealing with **certain situations.** Generally it is discussed as an aspect of religion. Religion is found in all established form of activities. It may include prayers. Military organization and other formally organized groups have adhered to a prescribed form of behavior known as **ritualism.**

Examples of Rituals

- ☆ Playing with crackers on 'Diwali'
- ☆ Celebration of Independence Day
- ☆ Celebration of Republic day

Conventions

These are customs regulating more significant social behavior. Parents generally do not care to leave such learning to chance. Parents instruct their children the conventions though often they (parents) cannot explain why the child must confirm.

Examples of Conventions

- ☆ Being polite to others
- ☆ Wearing clothes in public
- ☆ Using knife, spoon or fork for eating *etc.*

Differences between Mores and Taboos

Mores	Taboos
Mores refer to positive action	Taboos refer to negative action
Mores are the customs regarded by the forbidden members of the society	They are the customs which are as vital or essential
Things ought to be done	Things ought not to be done
e.g. Monogamy, honesty *etc.*	*e.g.* eating of beef in Hindu religion *etc.*

Differences between Mores and Folkways

Mores	Folkways
These are socially acceptable ways of	These are the customary ways of behavior that involve moral standards behaving in society
These are rigidly enforced and if not followed by a person the individual gets severe penalty form the society	Persons who do not conform may be subjected to criticism or be considered 'strange' but would not necessarily penalized

Mores	Folkways
Patterns of behavior which are considered essential by the society	Expected form of behavior but not rigidly enforced
If violated the group or society may be disturbed or divided	If violated will not have severe effect on society
e.g. Monogamy, honesty *etc.*	*e.g.* Good manners, greeting others *etc.*

Formal Social control

Formal social control refers to written rules and regulations that are deliberately created and enforced by authorized agencies. Important means of formal social control are laws, education and coercion.

1. Laws refer to a body of rules and regulations enacted and enforced by authorized agencies. These have become important as the human society has grown in size and complexity with predominance of secondary relationships. Laws define and regulate human behaviours. Their violation results in punishment or penalty as indicated in them. Consequently, social order and harmony are ensured for smooth living of people.

2. Education is a process of learning or gaining knowledge about the universe. Education helps in acquiring the patterns of living, values of discipline, co-operation, tolerance, positive attitudes, and sense of judgment as well as rational thinking among people essential for smooth social life.

3. Coercion is the use of force to achieve social order. It may be violent or non-violent in nature. This is used as means of social control when all other normal means fail to bring social order. Violent coercion may be in the form of physical injury, imprisonment, death punishment, *etc.* Non-violent coercion involves strikes, boycott, non-co-operation, *etc.*

Social Interactions

Human being by nature and necessity a social animal. As a social animal he meets with other human beings, interacts with them and establishes social relationship. Thus, when individuals and groups meet and establish social relationships they interact with each other. Such interaction is known as social interaction. This interaction can take place between two or more individuals or groups.

Definitions

Social interaction is dynamic interplay of forces in which contact between persons and groups results in a modification of the attitude and behaviour of the participants.

Gillin and Gillin: "By social process we mean those ways of interacting which we can observe when individuals and groups meet and establish system of relationships or what happens when changes disturb already existing modes of life."

Ginsberg: "Social processes mean the various modes of interaction between individuals or groups including co-operation and conflict, social differentiation and integration, development, arrest and decay".

Horton and Hunt: "The term Social process refers to the repetitive form of behaviour which is commonly found in social Life".

Aspects of Social Interaction

Four important aspects of social interaction have been identified

1. **Social contact:** Without social contact interaction is not possible. Social contact, as distinct from physical contact, exist when there is reciprocal response and an inner adjustment of behaviour to the actions of others

2. **Communication:** Communication is essential for interaction.

3. **Social structure:** The context for social interaction is the structure of society *i.e.* rural and urban.

4. **Forms of interaction:** This social interaction usually takes place in the form of

   a. Co-operation

   b. Competition

   c. Conflict

   d. Accommodation and

   e. Assimilation

These forms of social interaction are also called as social process. The social processes are the fundamental ways in which people interact and establish social relationships. The social process are classified in to conjunctive and disjunctive processes. **Conjunctive processes** refer to those patterns of social interaction which result in positive interaction. *e.g.* Co-operation, accommodation and assimilation. The **disjunctive process** are those which have an opposite effect so that peoples are pushed apart with less solidarity as in conflict and competition.

Co-operation

Cooperation means working together toward common objectives or goals. The word is derived from two Latin words-"Co" meaning together and "operate" meaning to work. Thus where two or more individuals or groups work or act together jointly in pursuit of a common objective. Cooperation may be brought about by several motivating factors and by situations involving such factors. The more important motivating factors are:

☆ Personal gain – Getting personal benefit through cooperative effect

☆ Common purpose- to solve common purpose – Village road

☆ Altruistic factors – to help others when they in crisis situation *e.g.* Helping people those who are affected by natural calamities.

☆ Situational necessity – during emergencies cooperative action is necessary.

Forms or Types of Co-operation

1. Cooperation that results from loyalty or adherence to the same objective.
2. Antagonistic cooperation often occurring in labour disputes when management and labour agree to discuss differences. Even with opposing objectives both sides are mutually dependent on one another and hence realise the necessity of working out cooperatively a mutually acceptable arrangement.
3. Cooperation that results from mutual dependence.
4. Cooperation that results from efforts to compete with others in order to achieve the goal before them.
5. Cooperation that is enforced as a result of subordination.
6. In addition to the above there are three other types of cooperation based on differences in group attitudes and group organisation.

   a. **Primary co-operation:** The group and individual fuse so that the group engulfs all or nearly all of the individuals life. Identification of individuals, group and task to be performed are interlinked. *e.g.* Daily routine work

   b. **Secondary cooperation:** Such cooperation, characteristics of modern western society, is highly formalised and specialised and occupies only part of an individuals life. *e.g.* Business life

   c. **Tertiary cooperation:** Tertiary cooperation is unsecure because of latent conflict underlying it. Two antagonistic political parties may cooperate to oppose a third party. Once the party is defeated their cooperation may cease.

Characteristics of Co-operation

1. Co-operation is a conscious process.
2. Co-operation is a continuous process.
3. Co-operation is a universal process.
4. Co-operation is a personal process.
5. Co-operation is an associative process of social interaction which takes place between two or more individuals.

Competition

Competition is the social process or form of social interaction in which two or more individuals or groups strive against each other for the possession or use of some material or non material good. The focus is primarily on the achievement of the objective desired by both and secondarily on each other. The goal or objective by its nature quality or quantity may be such that only one can achieve, or secure it, making the competition more intense.

According to Bogardus, "Competition is a contest to obtain something which does not exist in a quantity sufficient to meet the demand."

Mazumdar defines competition "as the impersonalized struggle among resembling creatures for goods and services which are scarce or limited in quantity."

Forms of Competition

Several sub types of competition have been identified. Some of more important are as follow.

1. Absolute and Relative Competition

Absolute competition exists when the goal is such it can be achieved or secured by one competitor only at a time, and he is declared the victor. *e.g.* Winning foot ball team. All the competitors must be eliminated before one can claim victory. Relative competition- on the other hand it is based on the degree to which a goal or objective may be achieved by competitors. *e.g.* Competition for money, for other forms of wealth or for prestige.

2. Personal and Impersonal Competition

In personal competition the focus of attention of each competitor is on other competitors whom he strives to eliminate as well as on the goal. Such personal competition often approaches conflict with rather a narrow dividing line. Impersonal competition on the other hand has no personal focus on individual rivals striving instead to reach a goal rather than to defeat an opponent. *e.g.* In a labour management dispute each side competes over wages – labour seeking the maximum and management the minimum.

Characteristics

1. Competition is impersonal struggle.
2. Competition is an unconscious activity.
3. Competition is universal.
4. Competition is a cause of social change.
5. Competition may be constructive or destructive.
6. Competition is continuous.
7. Competition is dynamic.
8. Competition is always governed by norms.
9. Competition is the source of motivation for the individuals.

Value of Social Functions of Competition

Competition performs useful functions in a society.

1. Competition determines the functions of individuals.
2. Competition is conducive to economic as well as social progress.
3. Competition is a source of motivation.
4. Competition provides better opportunity to satisfy their desire.

According to H.T. Mazumdar, competition performs five positive functions:

1. It determines the status and location of individual members in a system of hierarchy.
2. It aims to stimulate economy, efficiency and inventiveness.
3. It tends to enhance one's ego.
4. It prevents undue concentration of power.
5. It creates respect for the rules of the game.

Conflict

Conflict refers to the struggle in which competing parties attempting to reach a goal, strive to eliminate an opponent by making the other party ineffectual. Victory is at the expense of the opposing party.

Conflict has been defined as the process of seeking to monopolise rewards by eliminating or weakening the competitors.- **Horton**

According to A.W. **Green,** "the deliberate attempt to oppose coerce or force the will of another or others."

According to **Young and Mack,** "Conflict lakes the form of emotionalised and violent opposition in which the major concern is to overcome the opponent as a means of securing a given goal or reward."

Conflict Vs competition

Conflict	Competition
Chiefly in the focus and manner of achieving the goal	Primary focus is goal
Conflict involves contact	Does not involve contact
Takes place on a conscious level	Competition is unconscious
Conflict is an intermittent process	Competition is a continuous process
Conflict may involve violence	Competition is non violent
Conflict disregards social norms	Competition does care for these norms.

Characteristics

1. Conflict is affected by the nature of the group.
2. Frustration and insecurity promote conflict.
3. Conflict is always conscious.
4. Conflict is personal.
5. Conflict is not continuous but occasional.
6. Conflict is universal.
7. Conflict may be personal or impersonal.

Role of Conflict

H. T. Mazumdar mentions the following role of conflict:

1. Conflict leads to redemption of value system.
2. Conflict may lead to change in the relative status of conflicting parties.
3. Conflict may lead to new consensus.
4. Conflict tends to still in the morale and promote the solidarity of the group.
5. Conflict may lead to change in the relative status of conflicting parties.
6. Conflict, concluded with victory, leads to the enlargement of the victor group.

Accommodation

Accommodation is a process of developing temporary working agreements between individuals or group. – Horton.

Accommodation refers to a permanent or temporary termination of conflict parties to function together without open hostility at least in some respects- Cuber

The conflicting parties arrange for alternative to remove the conflict relationship and to enable some form of cooperation. Thus accommodation may be viewed both as a process of social interaction as well as the result of social interaction and it is one of the important and inevitable outcomes of a social situation of competition or conflict.

Forms of Accommodation

Accommodation may be achieved in various ways and take several forms.

1. **Compromise:** Each antagonistic party agrees to make concessions that allow them to reach an agreement. This give and take continuous until all parties are satisfied. *e.g.* Labour management

2. **Conversion:** One of the interacting parties accepts and adopts the views of the other. Conversion is frequently related to religious beliefs. Those who accept and adopt the religious beliefs and views of others are referred to as converts.

3. **Tolerance:** In this form of accommodation, interacting parties agree to disagree. Each party holds its own position, but respects the fact that the other party has an opposing view point. They tolerate each other, despite the fact that the basic issue is not eliminated.

4. **Arbitration:** When contending parties do not settle differences among themselves, arbitration is frequently employed- the problems is submitted to a mutually agreeable third party who act as a mediator, capable of studying the issue objectively. This procedure is frequently followed in reaching a compromise.

5. **Truce:** A truce is an agreement to cease rivalrous interaction for a definite or indefinite period of time. The purpose is usually to give both parties time

to review the issue in the light of proposals or suggestions for settlement. Illustration of the use of this form of accommodation are plentiful in various battles and wars.

6. **Subordination and superordination:** Subordination as a form of accommodation serves to structure relationships between a victor and the conquered at the end of a conflict. Accommodation by subordination is effective under two conditions. The first is that the dominant party be so strong as to force that other to submit. The second condition under which subordination as a form of accommodation may be successful is that relationships of subordination be socially sanctioned as a part of the social structure and heritage of the society.

7. **Displacement:** Displacement involves termination of one conflict by replacing it with another. For instance the threat of war may unify parties within a country.

8. **Institutionalised safety values:** The structure of various societies may provide institutionalised means for release of tensions, which may serve as a form of accommodation.

Characteristics of Accommodation

1. Accommodation is the result of conflict. If there were no conflict there would be no accommodation.
2. It is a mixture of both love and hatred.
3. It is a universal process.
4. It is a continuous process.

Assimilation

Assimilation has been referred to as the fusing and blending process whereby cultural difference tend to disappear and individuals and groups once dissimilar become similar.

Assimilation is the process of mutual cultural diffusion through which persons and groups become culturally alike- Horton.

According to **E. A. Bogardus,** "Assimilation is a process whereby attitude of many individuals are united and thus developed into a united group."

This process takes place when two different cultures meet, with the dominant culture assimilating the other.

Characteristics of Assimilation

1. Assimilation is a universal process.
2. It is a slow and gradual process.
3. It is ah associative process which is closely related to accommodation.
4. It is a cultural and psychological process.

5. It is also an unconscious process.

6. It is not a simple but a complex process.

Role and Importance of Assimilation

Like socialization, assimilation is a process of learning, but it starts when the individual comes in contact with other cultures. Assimilation is a social and psychological process. It is a result rather than a process. The social contacts thus established finally result in assimilation. The speed of the process of assimilation depends on the nature of the contacts. The ancient culture of India provides a number of examples of assimilation. The Aryans assimilated the Dravidian ideas. The Hindus and the Muslims in India through their prolonged living assimilated the culture of each other.

Forms of Assimilation

The three important kinds of assimilation are

1. A socialised individual in one culture may later move to another culture. In course of time he becomes assimilated into this second culture.

2. Two cultures merge into a third culture which, while somewhat distinct, has features of both merging cultures.

3. In small groups even in the family between husband and wife –assimilation may bring about a similarity of behaviour.

Chapter 11

Social Change

Change is the law of nature. The word 'change' denotes a difference in anything observed over some period of time. Therefore, social change means observable differences in any social phenomena over a period of time.

Definitions

"Social change is a term used to describe variations in or modifications of any aspect of social processes, social patterns, social interaction or social organization."

– Jones

"Social change is meant only such alterations as occur in social organization, that is, structure and functions of society."

– K. Davis

Social change is change in the relationships.

– Mac Iver

Social change results in both positive and negative modifications of and alternations in human life or society.

Dimensions

There are three important dimensions of social change; they are:

1. Structural change.
2. Functional change
3. Cultural change.

1. Structural change refer to changes in the structural forms of the society involves changes in roles, changes in class and caste structures, changes in the forms of social institutions such as the family, the Government, the educational system, *etc.*

2. Functional change or interactional dimensions refers to change in the social relationships in society between persons and groups. Changes may be in the frequency of contacts, a shift from primary relationships to secondary relationships *i.e.,* informal, personal and intimate to formal and contractual relationships, co-operative to competitive relationships, changes in the functions of social institutions like family, Government, Educational *etc.*

3. Cultural change refers to changes in the culture of society through discoveries and inventions, diffusion and borrowings of new technology.

Factors

There are four important factors of social change; they are:

1. Physical (Geographical) factors
2. Psychological factors
3. Cultural factors
4. Technological factors.

Physical or Geographical Factors

Such as climate, flood, famine, earthquake, *etc.* influence human life. Some times, they result in ill-effects. To control these effects, new technologies have been evolved. For example, dry land agriculture technologies have been evolved to grow certain crops in areas with less moisture; famine resistant crops are also evolved; new technology has been evolved for indicating the occurrence of earthquake or tsunami and also to control their adverse effects. Recurrent famines have also made the village people not only poor and jobless but also to migrate to other places where better opportunities of life are available. Consequently, the rate of rural migration has increased over years which have led to scarcity of labour for agricultural activities.

Psychological Factors

Like willingness, motivation, perception, outlook, progressive thinking contribute to the changes in human life or society.

Cultural Factors

Include education, social legislations, discoveries, inventions, diffusions and borrowings significantly bring about changes in society.

Education brings changes in human life by means of new ideas, values, skills, social patterns and rational thinking.

Social legislations have been responsible for eliminating many evil social practices and ensuring the right paths of living.

Cultural diffusion helps in spreading the patterns of living or technology of one group or society to the other by means of contact or influence of communication.

Technological Factors

Science and technology are the two important aspects of human society. Technology refers to the application of scientific knowledge to improve living conditions of people. Technology has been in existence in varied forms in human society since its inception. However, it has undergone changes due to more of inventions and discoveries. The important areas of modern technology which have brought about changes in human society are: (a) Industrialization, (b) Improved means of transport and communication, (c) Development of agriculture.

(a) Industrialization

Industrialization occurred in the middle of 18th century in England and later entered into other nations. This has brought about changes in production process through large scale factories. This system provided more employment opportunities to the people and at the same time adversely affected the village and cottage industries upon which the rural artisans had depended very much. Consequently, the rural artisans lost their jobs and moved in search of jobs and settled down in and around factories which led to emergence of urban centers (townships). As these town ships bulged in size, new infrastructural facilities were created to meet the needs of the people. This attracted more people from rural areas and thus the rate of rural migration increased over years.

(b) Improved means of Transport and Communication

Improved means of transport includes road, rail and air transport which helped people to move out of their community to other far-off places within a short span of time for various purposes. Consequently, the physical barriers that existed among people of different places are broken. This has helped to increase awareness among people about the happenings in and around their places. Further, cosmopolitan nature has developed among people due to their frequent visits to other places consequent to the improved means of transport.

Improved means of communication like television, radio, telephone, cell phones, computers, *etc.*, have helped people to establish contacts with others who are at a greater distance within a short span of time. Further, communication has also helped people in acquiring and adopting the culture of others. It has also helped in providing recreation and education to people and thus their outlook and horizon are broadened.

(c) Development of Agriculture

Agriculture development has taken place due to use of modern agriculture technology such as high yielding varieties of seeds, chemical fertilizers, pesticides and agricultural machinery. This has helped in increasing agricultural production and thus solved the food problem. Consequently, rural incomes have increased resulting in improvement of standard of living of rural people. The agricultural development is found to be more advantageous to the farmers particularly the large farmers in irrigated areas as against the farmers in rain fed area. Thus, social cleavage has widened among different categories of farmers. The use of agricultural machinery such as tractors, power tillers *etc.* has taken away the jobs of agricultural labourers who moved to the cities in search of jobs. This has resulted in increase in the rate of rural migration which creating scarcity of labourers for agriculture. As the modern agricultural technology is cost intensive, many new financial organizations like credit co-operative societies, rural banks, *etc.* have been established to help the farmers. The over use of chemical fertilizers and pesticides has resulted in health hazards, environmental pollution and hence efforts have now been made to control the use of these by means of organic farming which is believed to be harmless.

Indicators/Dimensions of Socio-economic Changes

The folowing are the dimensions of socio-economic changes

1. Education
2. Health maintenance and Nutrition
3. Shelter/Housing
4. Occupation
5. Cultural factors
6. Out side contact/communication
7. Economic changes
8. Social Participation
9. Women Status
10. Material possession
11. Migration
12. Over all Facility changes.

Chapter 11

Leader and Leadership

Definitions of leader

Leader is a person who exerts an influence over a number of people.

Leader is one who leads by initiation of social behaviour, by directing, organizing or controlling the efforts of others, by prestige or power or position.

Leader is a person who is spontaneously considered or chosen as influential in a given situation. In every society certain individuals operate within groups to guide and influence members to action. These individuals are referred as leaders.

Definitions of Leadership

Leadership is defined as an activity in which effort is made to influence people to cooperate in achieving a goal viewed by the group as desirable.

– Rogers and Olmsted

Leadership is defined as the role and status of one or more individuals in the structure and functioning of group organizations, which enable these groups to meet a need or purpose that can be achieved only through the co-operation of the members of the group.

– Hepple

Classification of Leadership or Types of Leaders

There are several classifications of leaders. For example the leaders may be classified in terms of the types of groups they work with such as political, military, business, religious, recreational leaders *etc*. Whyte has classified leaders in to 4 categories as follows:

1. **Operational leaders**: those persons who actually initiate action within the group, regardless of whether or not they hold an elected office

2. **Popularity leaders**: means in a group a popular person will be elected to a position of leadership because the members like him. Sometimes such an individual may or may not be the actual leader of the group. Such persons holding elective positions do very little about initiating action for the group and are mere figureheads or ornamental leaders. They are also called nominal leaders

3. **Assumed representative type**: refers to a person selected to work with a committee or other leaders because the latter (Group B) have assumed that he represents another group (Group A) they desire to work with; he may or may not be a leader of the group

4. **Prominent talent**: *e.g.* artists and musicians who have exhibited an outstanding ability and accomplishment in their respective fields. It may include the experts and intellectual leaders

Another classification divides leaders in to 2 categories:

1. **Professional leaders**: the professional leader is one who has received specific specialized training in the field. He works full time as an occupation and is paid for his work. E. G. Extension Officer, Gram Sevak, Agricultural Officer *etc.*

2. **Lay leaders**: the lay leader may or may not have received special training, is not paid for his work and usually works part time *e.g.* youth club president, Gram Sahayak *etc.* Lay leaders also called as Volunteer leaders, or local leaders or natural leaders. These local leaders may be either formal leaders or informal leaders, depending on whether they are regular office bearers of organized groups or not.

Perhaps the most significant classification form the viewpoint of modern research as well as practical application of the results of research is the one designating them into the following **three** types:

1. Autocratic Leader

Autocratic leader is also known as authoritarian leader. He operates as if he cannot trust people. He thinks his subordinates are never doing what they should do; that the employee is paid to work and therefore must work. If he is a benevolent (kind) autocrat he may tend to view employees as children and encourage them to come to him with all their problems, no matter what is the nature or magnitude of the problem. The results of his leadership are

a. Most employees develop a sense of frustration, and finally feel insecure in their job

b. Work slows down or stops completely when the supervisor is away

c. The employee's needs for a feeling of importance and satisfaction are not met

    d. Employees are kept dependent on the supervisor; thus they have no opportunity to show initiative

    e. Employees frequently either become aggressive or alternatively identify closely with supervisor (submissive yes-men)

2. Democratic Leader

He shares with the group members the decision making and planning of activities. The participation of all members is encouraged. He works to develop a feeling of responsibility on the part of every member of the group. He attempts to understand the position and feelings of the employee. If he criticizes, he does so in terms of results expected, rather than on the basis of personalities. The results of his leadership are:

    a. Employees produce a larger quantity and higher quality of work

    b. Individual and group morale are high

    c. Employee's basic needs to participate and feel important are met

    d. Employees feel secure

    e. Employees seldom become aggressive

    f. The supervisor finds that less supervision is necessary

3. Laissez-faire Leader

He believes that if you leave workers alone, the work will be done. He seems to have no confidence in himself. If at all possible he puts off decision-making. He tends to withdraw from the work group. He is often a rationalizer. The results of his leadership are:

    a. Low morale and low productivity within the work group

    b. Employees are restless and lack incentive of 'team work'

    c. Another leader often an informal leader arises

    d. Problems of administration supervision, and coordination. Symptoms of disorder 'anarchy' are seen

Roles of Leader in a Group

Groups are dependent on leaders. A leader is not only a member of group and also is the focal point of activity of his group. He plays an important role in group's activity. The important roles of the leader are as follows:

1. **Group initiator:** the most important role of leader is that he should take initiative to get the group in to action

2. **Group spokesman:** if the group is to have outside relations it must be able to speak as a unit and leader is its voice. Leader has the responsibility of speaking for the group and representing the interests of the group

3. **Group harmonizer:** in all groups uniformities and differences are formed. A leader should be able to resolve differences peacefully. The role of the

group harmonizer is to promote harmony in the group in line with basic purpose of the group

4. **Group planner:** generally it is assumed that the person chosen for leadership know a little bit more about the problems which the group is facing and the possible solutions. So the leader has to plan the way by which the group can satisfy its needs. The leader has to plan for the group and with the group

5. **Group executive:** the leader is one who takes important role in conducting business of the group and he is responsible for seeing that the business of the organization is carried on according to democratic principles. It is the job of the leader that individuals of group accept responsibility of their part of activities in any plan of action adopted by the group

6. **Group educator or teacher:** in most of the groups the leader will have more training and experience. So the leader can teach according to the level of understanding of the members of the group so that they can understand his views. In this capacity his chief function is to develop and train other leaders so that group is not dependent completely on him

7. **Group symbol or symbol of group ideas:** all social groups have implicit (internal) or explicit (external) norms or ideals. As a rule persons accepted as leaders are those who have adopted these norms or ideals and live by them. The leader must make the members feel that they need ideals and depend upon them for accomplishing what they desire to do, the leader should be not be self interested

8. **Group supervisor:** the leader also acts as supervisor. A good leader supervises the work of his peers and subordinates. Professional leaders such as Extension Officers, in addition to serving as leaders of social groups also devote a portion of their time to working with lay leaders and group organizations like youth clubs, cooperatives *etc.*

Qualities or Traits of a Leader

Hepple has listed the following traits or qualities are desirable for effective leadership.

1. Physical fitness
2. Mental ability
3. Sense of purpose
4. Social insight (Sensitivity to other persons position, problems or points of view)
5. Communication
6. Love for people (friendliness without favouritism)
7. Democracy (Giving all members equal opportunities for participation)
8. Initiative
9. Enthusiasm

10. Authority (mastery of knowledge and skills in a particular field)
11. Decisiveness
12. Integrity or character
13. Teaching ability
14. Convictions and faith

Leadership Style

A leadership style is a leader's style of providing direction, implementing plans, and motivating people. There are many different leadership styles that can be exhibited by leaders in the political, business or other fields.

1. **Authoritarian**
2. **Paternalistic Leadership**: In this style of leadership the leader supplies complete concern for his followers or workers. In return he receives the complete trust and loyalty of his people. Workers under this style of leader are expected to become totally committed to what the leader believes and will not strive off and work independently. The relationship between these co-workers and leader are extremely solid. The workers are expected to stay with a company for a longer period of time because of the loyalty and trust. Not only do they treat each other like family inside the work force, but outside too. These workers are able to go to each other with any problems they have regarding something because they believe in what they say is going to truly help them.
3. **Democratic**
4. **Laissez-faire**: This is an effective style to use when:
   ☆ Followers are highly skilled, experienced, and educated.
   ☆ Followers have pride in their work and the drive to do it successfully on their own.
   ☆ Outside experts, such as staff specialists or consultants are being used.
   ☆ Followers are trustworthy and experienced.
   This style should NOT be used when:
   ☆ Followers feel insecure at the unavailability of a leader.
   ☆ The leader cannot or will not provide regular feedback to their followers.
5. **Transactional**: The transactional style of leadership. Mainly used by management, transactional leaders focus their leadership on motivating followers through a system of rewards and punishments. There are two factors which form the basis for this system, Contingent Reward and management-by-exception.
   ☆ **Contingent Reward:** Provides rewards, materialistic or psychological, for effort and recognizes good performance.

☆ **Management-by-Exception:** Allows the leader to maintain the status quo. The leader intervenes when subordinates do not meet acceptable performance levels and initiates corrective action to improve performance.

6. **Transformational**: A transformational leader is a type of person in which the leader is not limited by his or her followers' perception. The main objective is to work to *change* or *transform* his or her followers' *needs* and *redirect* their thinking. Leaders that follow the transformation style of leading, challenge and inspire their followers with a sense of purpose and excitement. They also create a vision of what they aspire to be, and communicate this idea to others (their followers).

Different Methods of Selection of both Professional and Lay Leaders

Selection of Professional Leaders

A. Interview

1. The time-honored and most widely used method of selecting persons for position of professional leadership. It is based primarily upon an interview and an evaluation of past academic and occupational records of the individual. A large amount of information concerning a person can be acquired through an interview

2. The chief difficulty with the interview is that one can observe and evaluate the applicant only as he answers questions during a brief period of time

3. In industry and management there has been an attempt to supplement the interview by subjecting applicants to a battery of tests

4. These tests measure ability, aptitudes, attitudes and interests and both the academic training and practical experience

5. The use of a battery of tests along with an interview provides a better basis for selection than using the interview alone

B. Performance Tests

1. These have been used in certain situations as a part of the basis for selection of professional leaders

2. One type of these is the 'Leaderless group tests' in which seven or eight persons are given a common task to perform and it is left up to the persons involved to determine which person have become the leader

3. Another type of test is to appoint an individual as a leader and then observe how well he directs the activities of the members of the group

4. The big advantage of these performance tests is that one can observe the potential leader in a real life situation in which he is functioning as the leader of a group

Selection of Lay Leaders

A. Sociometry

1. Sociometry is concerned primarily with obtaining choices in inter-personal relations, such as with whom one would like to work, play *etc.* or to whom one would go for advice on farming or other problems

2. It attempts to describe social phenomena in quantitative terms

3. It may be used in selecting professional leaders also, but of greater use in selection of lay leaders

4. It is necessary that all the persons involved in a sociometric test know one another. These tests are not designed to measure vague factor called popularity, but it is popularity of acceptance in terms of specific activities.

5. Sociograms for the same individuals will manifest (bring out) differences when the choices are in relation to different activities. This method is very useful to the extension worker in finding out the natural or local or informal leaders in the villages. An extension worker goes into a given area and asks the farmers to indicate whom; they ordinarily consult for advice on farming, which the extension worker wants to introduce. Usually after a few interviews, it becomes apparent (clear) which farmer is the influential person on natural leader. The figure below illustrates the Sociometry test.

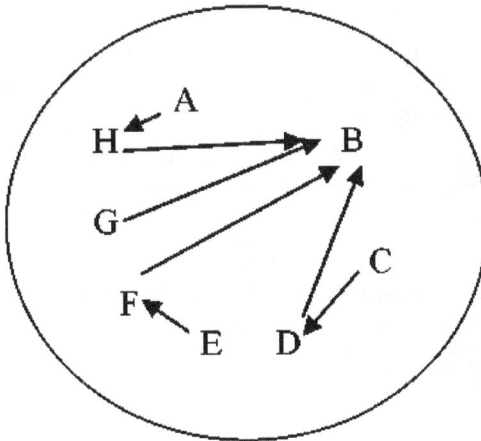

Sociogram

When farmer H is interviewed he may indicate that generally he goes to B for advice on farming, farmers G, F, D may also say that they take advice from farmer B on farming. The farmers A, E and C are depending on farmers H, F and D respectively. Then B is the operational or potential natural leader for these farmers and therefore if extension worker induces farmer B for the adoption of new improved practices it is quiet likely that the other farmers will be influenced by his behavior and adopt the same practices

B. Election

1. Another method widely used in selecting leaders, consists simply of the members of the group electing a leader through voting or any other method

2. The extension worker can guide or assist the local people in electing the right person for the right job by explaining to the group, the functions of leader in relation to particular problem and outlining the qualifications of a good leader for the given purpose. Election can also be used for selecting persons to receive leadership training who later become the actual leaders

C. The Discussion Method

1. Through discussions (on any subject) the person with sound knowledge and ability is soon recognized and a mere talker easily spotted

2. Discussion gives encouragement and assurance to the potential leader to express himself and over a period of time may make him more confident in accepting some position of leadership and he emerges as a valuable leader

D. The Workshop Method

1. In this method a large group is broken in to **smaller groups** and the responsibility of the program and decision-making rests upon the smaller units

2. Leadership emerges in each small group. Over a period of time, the extension worker can spot certain leaders who come to the fore (front) in taking responsibilities

3. The extension worker or professional leader in the workshop has the position of consultant, observer, discussion group leader *etc.*

E. The Group Observer

1. The extension worker should watch (observe) a community or group in action and then he will be able to spot potential leaders

2. He may observe the community in any type of situation. For obtaining the best results, the group should not be aware of that the extension worker is observing them

   Rogers who designated the local leaders as **opinion leaders** mentions the following two methods to locate these leaders in mass public

F. Key Informants

1. In a community **key informants** or persons with important information about their community like teachers, VLWs etc may be asked by the extension worker to indicate opinion leaders in that area based on their indications he will select the leader.

2. Key informant method is cost saving and time saving when compared to the sociometric method and other methods

G. Self-designating Technique

This consists of asking a respondent a series of questions to determine the degree to which he perceives **himself** to be an opinion leader based on the analysis of the answers obtained, the extension workers selects a leader

Lay leaders are otherwise called as local leaders or **informal leaders** or **volunteer leaders**. Professional leaders are otherwise called as **formal leaders.**

Chapter 12

Education – Psychology – Educational Psychology – Social Psychology

The term psychology was derived from Greek; 'psyche' meaning is soul and 'logos' means Science. Hence about 2500 years back it was referred to a 'science of soul. In olden days, it was believed that soul was responsible for various activities of man such as thinking, imagining, reasoning *etc.*

In the Middle Ages psychology became a 'science of mind'. Then in 20th century psychology assumed scientific look and it became the 'science of mental behaviour'. By observing one's behaviour we can have knowledge of one's conscious and unconscious minds.

Father of Psychology is **Sigmund Freud.**

Meaning of Psychology

An individual's behaviour consists not only of his observable act but also all his reactions to inner states and to environmental factors of influence. The human organism is extremely complex. The environmental factors that can affect the organism include all the persons, situations and conditions that- constitute the external world of any living individual.

Psychology is concerned with discovering the ways in which individuals and groups at different age levels, tends to respond to environmental stimuli. According to data obtained from scientifically conducted studies of human behaviour, it has been concluded that people tend to react similarly in certain situations and under certain conditions.

Psychologists are interested in why as well as what similarities and differences among human reactions. Various schools of thoughts have arisen. Most of the pertinent assumptions are:

a. Human behaviour is functional and dynamic.

b. At every stage of development, an individual's reactions are influenced by the effect of his experiences with people, things, situations and environmental conditions upon his desire to satisfy felt needs, wants and urges.

c. A person's reactions usually represent the functioning of a total integrated pattern of behaviour.

Psychology is also concerned with conscious and unconscious behaviour, and normal as well as abnormal behaviour.

Definition of Psychology

1. Psychology is the science of human and animal behaviour.

2. Psychology is a field of study which seeks scientific methods to describe, understand, predict and control the behaviour of living organisms.

Importance of Psychology

(Importance of Psychology in Agril. Extension)

The study of psychology as the science of human behavior helps in identifying

1. The **abilities** of individual

2. The **needs** of individual and techniques to be employed to **motivate** them

3. The **hereditary** and **environmental** factors the affect the behavior

4. The levels of **achievement motivation** of the individuals

5. The factors that result in individual, intellectual **differences** and reasons for people becoming **problem men**

6. The factors that lead to **differential perceptions**

7. The causes of **retarded** learning

8. The causes of **emotions** and **frustration** in human beings

9. The causes of **forgetting** and how to improve **memory**

10. The levels of knowledge, attitudes possessed by the individuals

11. The different **psychological traits** possessed by individuals. By the application of different tests and help in evaluation of the behavior of the individual.

Branches of Psychology

Psychology now has several branches. The branches are:

1. **General psychology**: It is the study of the basic principles or all aspects of psychology. It is the basis for all other branches and applications.

2. **Abnormal psychology**: Abnormal psychology deals with the nature of unconscious mind.

3. **Clinical or medical psychology**: It is an applied branch which uses the general findings of abnormal psychology to the treatment of psychological disorders.

4. **Child psychology**: This branch deals with the various developmental processes of child.

5. **Animal or comparative psychology**.

6. **Para psychology**: It deals with problems like extrasensory.

7. **Experimental psychology**: It is constituted by the experimental studies of the various aspects of general psychology conducted under experimental conditions in the laboratory with the help, of tests, tools, techniques *etc.* Mostly animals are used in experiments.

8. **Industrial psychology:** It is the study of all mental processes and functions when human beings are made to work together in an organised enterprise. It also deals with selection of right person for a right job. Various aptitude tests have been developed in this branch. Advertising is also covered under this branch.

9. **Social psychology**: This branch is developed out of sociology and cultural anthropology. It is concerned with the study of the individual as a member of a group and relations of groups to one another. It also studies about the emergence of social behaviour.

10. **Educational psychology**: It is a branch of general psychology deals with the behaviour of human beings in educational situations. This means that it is concerned with the study of human behaviour or the human personality - its growth, development and guidance under the social process of education.

Educational psychology can be regarded as an applied science in that it seeks to explain learning according to scientifically determined principles and facts concerning human behaviour.

It deals with such topics as:

1. The learner characteristics, his methods of learning, *etc.*, that affect his learning efficiency.

2. The teacher characteristics, teaching methods and their impact on learners and their learning efficiency.

3. The influence of various social and situational factors that operate in the learning situation and their impact on learning efficiency.

4. The techniques of evaluation of learning outcomes in the learners.

Scope and its Importance in Agricultural Extension

1. Educational psychology studies the limitations and qualities of individuals - physical capacity, intelligence, aptitude, interests, *etc.* which play a major role in one's learning.

2. It helps in improving teaching and learning. This branch helps in formulating training programmes for improving the skill of teachers and methods for organizing good learning situations.

3. Psychology attempts to discover the source of knowledge, belief, customs and to trace the development of thinking and reasoning so as to find the kind of environmental stimulation that produces certain type of activity.

4. It will help extension workers to find causes of prejudices, the habit of sticking to old practices and ways of doing things, the doubts and lack of confidence and factors affecting motivation.

5. It also helps them to know the emotions and feelings of farmers, how villagers or farmers learn new practices and what type of approaches be adopted and teaching aids be used.

Definition of Social Psychology

Social psychology is defined as the branch of knowledge which studies the relationships arising out of the interaction of individuals with each other in social situations. In brief it deals with thinking, feeling, and acting of an individual in society.

Social psychology, as discussed earlier, attempts to determine the character of social behaviour. Social behaviour involves one of the four following basic reactions.

1. When one individual meets another individual there is reaction. Each individual affects the other individual with whom he comes into contact and is inturn affected by them.

2. Individual may be reacting to group (*e.g.*) extension worker meeting a group of farmers.

3. As a counterpart of the above situation there will be reaction of a group of individuals to a single individual (*e.g.*) group meeting its leader.

4. There is reaction of one group of individuals to another group of individuals.

Social psychology studies the characteristics of all these four forms of social behaviour. It must, however, be borne in mind that social psychology studies the individual and not the group itself. Social psychology studies the individual in relation to his fellow-men.

Chapter 13

Basic Principles of Human Behaviour – Sensation, Attention, Perception

Human Behaviour

It is the expression of one's thoughts and feelings. Behaviour patterns that are expressed outwardly are called as *overt behaviour patterns* and those that are internal are called as covert behaviour patterns.

Human behavior refers to the range of behaviors exhibited by humans and which are influenced by culture, attitudes, emotions, values, ethics, authority, rapport, hypnosis, persuasion, coercion and or genetics.

The behavior of people (and other organisms or even mechanisms) falls within a range with some behavior being common, some unusual, some acceptable, and some outside acceptable limits. In sociology, behavior in general is considered as having no meaning, being not directed at other people, and thus is the most basic human action. Behavior in this general sense should not be mistaken with social behavior, which is a more advanced action, as social behavior is behavior specifically directed at other people. The acceptability of behavior is evaluated relative to social norms and regulated by various means of social control.

Psychology studies mental behaviour. All activities or behaviour patterns could be fitted into stimulus-response mechanism.

Stimulus-Response Mechanism

A stimulus is anything that arouses the organism or any of its parts to activity

e.g. Light is the stimulus for eyes and sound for ears.

A response is any resultant activity that is aroused by a stimulus.

e.g. Somebody pricks you with a pin and you lift your arm. Here pin prick is the stimulus and lifting of arm is the response.

Type of stimulus response: External stimulus and External response.

Internal stimulus: Stimulus need not be external. It may sometimes be internal.

e.g. Feeling hungry and taking food

Internal response: Changes in one's organic or physiological conditions as a result of some stimulus.

All behaviour patterns and mental phenomena are due to this S-R mechanism.

The S-R mechanism involves the following things:

Stimulus ⟶ Receptor ⟶ Message ⟶ Connector ⟶ Effector ⟶ Result

1. **Stimulus:** The trigger, or cause of the change.
2. **Receptors:** Stimulus is converted to an electrical response

    Refer to sense organs possessed by organism through which stimuli are received. The environment of objects and persons influence these receptors and these receptors get impressions from objects and persons. Receptors are sensory cells specialized for sensitivity to environmental stimulation.
3. **Message:** Electrical impulse is sent to the CNS by sensory neurons. At the CNS, the impulse is sent to the brain via interneurons.
    a. Sensory to CNS
    b. At the CNS, impulse is relayed from sensory to interneuron
    c. Sensory message is sent to the brain
    d. The brain interprets the message
    e. The brain sends a response message to motor neurons via interneurons
4. **Connectors:** Those connect receptors and effectors. For example, The spinal cord is connected with receptors and also units the effectors. These nerves are called *sensory nerves* and also called as *in-carrying or off afferent nerves.*
5. **Effectors:** Response message is relayed onto motor neuron and sent to the effector.
6. **Result or response:** Activity based on message from the brain.

Central Nervous system

It is very important. It comprises of brain, spinal cord and all nerves emerging from these and running to the different parts of body.

Peripheral Nervous System

Nerves spread over different parts of the body and connecting them.

To some extent central nervous system and peripheral nervous system function independently. But most activities are largely under the control of central nervous system.

Neurons Types

1. Sensory neurons: Carry messages from the peripheral organs of the body to the central nervous system.
2. Motor neurons: Found in nerves carrying mes-sages from various centres of central nervous system to different peripheral parts of body.

Basic Principles of Behaviour

1. Sensations

Sensations are the simplest mental activities of man. A sensation is the awareness of a quality of an object that stimulates any sense organ.

e.g. Visual sensations of light and colour - related to eyes.

Auditory sensation of noise and tone – related to the ears.

2. Attention

Attention is the cognitive process of selectively concentrating on one aspect of the environment while ignoring other things. Attention has also been referred to as the allocation of processing resources. Examples include listening carefully to what someone is saying while ignoring other conversations in a room or listening to a cell phone conversation while driving a car. This is the process of attending to series of stimuli. From among the many stimuli which are within range psychologically we select only those that are related to our present needs and interests.

3. Perception

Perception is the process of understanding sensations or attaching meanings based on past experience to signs. For example, vision involves light striking the retinas of the eyes, smell is mediated by odor molecules and hearing involves pressure waves. Perception is not the passive receipt of these signals, but can be shaped by learning, memory and expectation

4. Attitude

Attitudes are generally positive or negative views of a person, place, thing, or event, activities, ideas, or just about anything in your environment

5. Motivation

Motivation is the process of initiating a conscious and purposeful action. Motive means an urge, or combination of urge to induce conscious or purposeful action. It is goal -directed.

6. Emotions

Emotion is a reaction involving subjective feelings, physiological response, cognitive interpretation and behavioural expression.

Emotions largely determine human behaviour and extension workers should learn how to utilise them for the purpose of education of rural people.

7. Frustration

In psychology, frustration is a common emotional response to opposition. Related to anger and disappointment, it arises from the perceived resistance to the fulfillment of individual will. The greater the obstruction, and the greater the will, the more the frustration is likely to be.

Sensations

Sensations are the gateway to knowledge. Sensations are the simplest mental activities of man. A sensation is the awareness of a quality of an object that stimulates any sense organ. There are as many kinds of sensations as there are sense organs. Each sensation is important and each has particular sense organs.

e.g. Visual sensations of light and colour - related to eyes.

Auditory sensation of noise and tone – related to the ears

It is defined as just awareness of stimulus. It is the sensory receptors and the transmission of sensory information to the central nervous system

Sensation is the function of the low-level biochemical and neurological events that begin with the impinging of a stimulus upon the receptor cells of a sensory organ. It is the detection of the elementary properties of a stimulus.

In other words, sensations are the first stages in the functioning of senses to represent stimuli from the environment, and perception is a higher brain function about interpreting events and objects in the world.

Types of Sensations

Sl.No.	Name of Sensation	Sense Organ Concerned	Types of Sensory Experience
1	Vision	The human eye, retina	Light, colour, shape, form etc.
2	Audition	The ear, the basilar membrane	Sounds of different types.
3	Taste (Gustatory)	The tongue- taste buds	Sweet, sour, bitter, spicy etc
4	Olfaction	The nose-receptors in the skin	Smells, fragrance, pungent etc
5	Cutaneous	The skin - receptors in the skin	Heat, cold, pain and pressure
6	Kinesthetic	The muscles - receptors in the muscles	Senses pull, push, strain etc
7	Organic	Receptors and the muscles of the internal organs	Bodily sensations like hunger, nausea, etc
8	Static or posture	Ear semicircular canals	Sense of equilibrium, dizziness, reeling etc.

Characteristics of Sensation

1. Quality

A visual sensation may be of one colour or another. The quality of a taste sensation may be sour, sweet, bitter *etc.*

2. Intensity

Each sensation may vary in intensity from low to high in a continuous manner. Thus, we experience mild pain or severe pain, faint light or bright light and so on.

3. Threshold

a) Absolute Threshold

For any stimulation to be aroused, the stimulus (light, sound, touch *etc.*) must have a minimum intensity. Stimuli of very low value are not responded to. Thus, for every sensation there is an absolute threshold level.

b) Differential Threshold

When we listen to some sound, we do not respond to every small change in the sound. Hence, changes in stimuli are not sensed unless the change involved is atleast of a certain minimal intensity. This change in the sensory experience is known as differential threshold value of stimulus.

4. Adaptation

A sensory system is able to adapt itself to a sensation if it is subjected to that sensation for a long time.

5. After-images or After-sensation

Strong sensory experiences continue to remain for a few more moments even after their sensation.

6. Extensity

Sensations may also vary in extensity or size, (eg.) Thus, we may see a small patch or large patch of light.

7. Duration

Sensations also possess the property of duration. Our sense experiences last for different lengths. The sensation may present for a long time or may disappear immediately.

8. Latency Period

Sensations have a latency period. This is the time taken by the bodily tissues before they start making their normal responses.

Attention

This is the process of attending to series of stimuli. From among the many stimuli which are within range psychologically we select only those that are related to our present needs and interests.

Objective Factors

The objective factors mean the factors possessed by the object, by which it attracts our attention. They are intensity, colour, size, repetition, movement, change, systematic form and novelty.

1. Intensity

Our attention is attracted by a louder sound than a weaker one. A bright colour attracts our attention than a lighter shade. Of two stimuli the stronger one has an advantage over the weaker.

2. Colour

A bright colour attracts our attention than a lighter shade.

3. Size

Bigger the stimulus better the chances of its catching our attention. Among a group of boys, the boy who is bigger in size than others will easily attract our attention.

4. Repetition

Sometimes, a stimulus even when it is not very intense or big in size, may still attract our attention by being repeated several times. If an advertisement is given several times repeatedly easily it will catch our attention.

5. Movement

A moving object will easily attract our attention.

6. Change

If there is a sudden change either in the intensity or in the size or even by way of sudden stopping then we attend at once to the change in the stimulus. For example, the ticking of clock in our room usually will not draw our attention because of its being repeated continuously. But if it suddenly stops we may at once turn and remark why has the clock stopped?

7. Systematic Form

We easily attend to those stimuli which have a definite systematic pattern. Some melodious tune though sung very softly may be pickedup for attention even in midst of other louder noises. Gestalt psychologists have emphasized the importance of the systematic form.

8. Novelty

Anything strange or unusual whether a sight, sound or any other sensation invariably draws our attention.

Subjective Factors

Subjective factors mean the factors possessed by individual, which facilitate his attention. They are interest, organic state, habits and inner drives.

- ☆ **Interest is an important Subjective factor:** One who is interested in gardening may be attracted by a new kind of flower or plant. He attends to it because of his own developed interest and knowledge.

- ☆ **Organic states also play an important role:** A hungry person may be attracted by anything that could be eaten in preference to other objects. Our mood also determines our attention, when we are in happy mood we may notice things that are pleasant.

- ☆ **Affection, motivation and inner drives also play great part in attracting our attention:** A sleeping mother may not be disturbed by a whole lot of loud noise outside. But if her sick child raises even a faint cry, it attracts her attention.

- ☆ **Habits:** Habits also can help in selection of stimuli, as children we are taught to attend to certain types of stimuli and neglect others and these habits of attention determine to a large extent the trend of our attention in later life

Shifting of Attention

Attention shifts from one thing to another very rapidly. We can attend to a thing continuously for a few seconds. Continuous attention means continuity with plenty of shifting. Now and then attention will be diverted but will be immediately brought back. The eyes do not steadily gaze at anything for any length of time beyond a few seconds. Such shifts occur mostly because of the fatigue affecting the sense receptors. Shifting of attention is also referred to as fluctuation of attention.

Span of Attention

Span of attention means how many letters or digits that we can see at a single glance. How many figures or letters can one notice in one act of attention? This can be determined by the use of an apparatus called the *'Tachistoscope'*. There are individual differences but usually 4 or 5 numbers or letters can be attended to at a single glance. The registration plate of a motor car contains usually only for 4 figures. Serial numbering will go upto 9999, but not 10,000. This is because when a car runs fast the traffic constable will not ordinarily be able to take note of more than 4 numbers.

Perception

Perception is the process of understanding sensations or attaching meanings based on past experience to signs.

Characteristics of Perception

☆ **Perception shifts:** Just like attention perception also shifts. As we attend to one part of the stimulus we perceive that part and then as attention passes on to another part we perceive-that part also.

☆ **Perception is a grouping and combining response:** We put several stimuli together and make a joint response to it. When we perceive the face of a friend there are several stimuli coming to us from different parts of his face eyes, ears, nose *etc.* We put them all together and understand it as a totality.

☆ **Figure has advantage over background in perception:** There are no gaps in nature and the human miqd also hates gaps. It tends to fill in gaps and perceive things as having a definite form.

☆ **Perception is an isolating response:** We perceive the thing we select for our attention and do not perceive the things that are not attended to.

☆ **Perception follows the 'Law of Reduced cues':** Applied "to perception, the law of reduced cues means that as we become more and more acquinted with an object, the signs by which we can perceive it become less and less till at last, a fraction of the original sign is enough for us to recognise that object.

Determinants of Perception

The various factors that determine our perception can be grouped as follows:

☆ **The sense organs:** Perception depends upon sense impression and upon the number, structure and function of the available sense organs. For example, if colours are not developed in the retina there cannot be perception of colour. Similarly absence of certain taste buds will limit one's taste perception.

☆ **Brain function:** Perception depends on the nature of the brain function. This gives us various frames of reference against which perception is made. Certain relations such as bigger and smaller, lighter and heavier, above and below *etc.*, are all perceived because of the function of the brain.

☆ **Past experience:** Perception also depends on one's past experience. The few light sensations that come from a ship are interpreted as a ship because of our past experience. We are able to supplement a number of characteristics that are not sensed at that particular moment. Past experience may also influence perception in the form of creating various kinds of prejudices and assumptions regarding the object perceived.

☆ **Set or attitude:** Perception also depends very much on one's set or attitude. This is the subjective condition.

☆ **Organic conditions:** One's organic condition will also influence his perception. The individual who is starving from hunger will easily perceive the eatable objects. One's motive also determines his perception.

Errors of Perception

There are several possibilities of our perceptions process being wrong and misleading. Such errors of perception are studied as two different phenomena *viz.,* illusions and hallucinations.

(a) Illusions

An illusion is a wrong or mistaken perception. The perceptual process always involves an interpretation of the sensory experience in the light of our past experience or present attitude, our organic needs *etc.* In some cases this interpretation is done wrongly and so the stimulus is perceived wrongly. Such a phenomenon is called 'illusion'. (*e.g.*: We perceive the coil of a rope in darkness as a snake.)

Psychologists have experimented with a number of geometrical designs to understand the phenomenon of illusion. Two of the well known examples.

1. Muller-Lyer-illusion
2. Horizontal-vertical illusion.

In the Muller-Lyer illusion there are two straight lines of equal length. One is bounded at the two ends by pairs of short opening outwards. The other is bounded by two pairs of short lines which' are reversed and give the idea of closure.

Though the two lines are equal in length invariably the latter is perceived to be shorter than the former. This is an illusion.

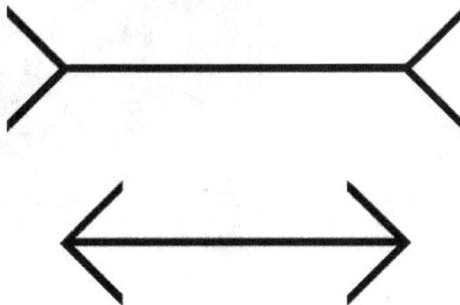

In the horizontal-vertical illusion there are two straight lines. One horizontal and other vertical. Both are of equal length. But invariably the vertical line is perceived to be longer than the other.

Both the Muller-Lyer and Horizontal-vertical illusions are optical illusions. We do have other illusions such as auditary, factual *etc.*

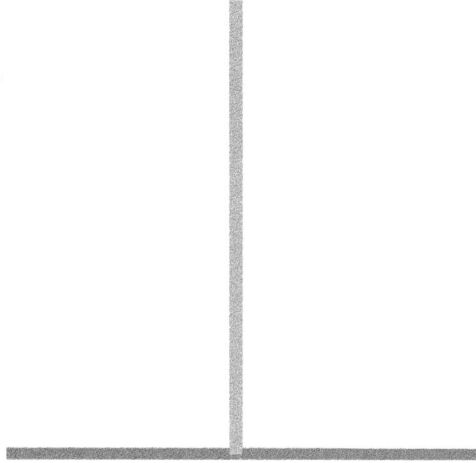

(b) Hallucinations

We perceive a figure or an object purely because of our subjective conditions, when there is no stimulus at all.

Such an error in perception which has no basis in a real sensory stimulus is called 'hallucination'. While illusion is wrong perception, hallucination is false perception.

If at night we see a ghost when there is practically no stimulus in the form of a human figure or anything resembling it would be an example of hallucination.

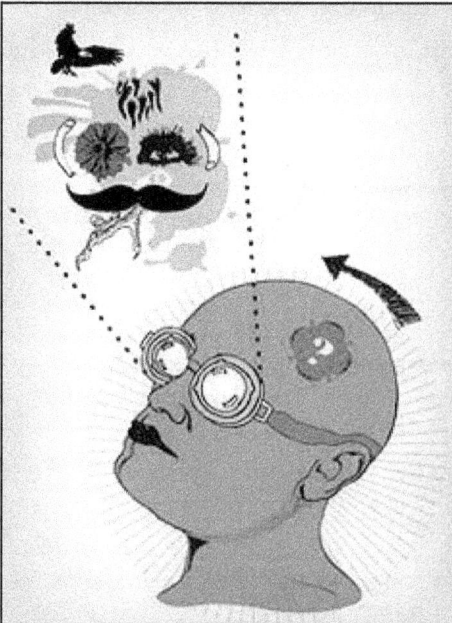

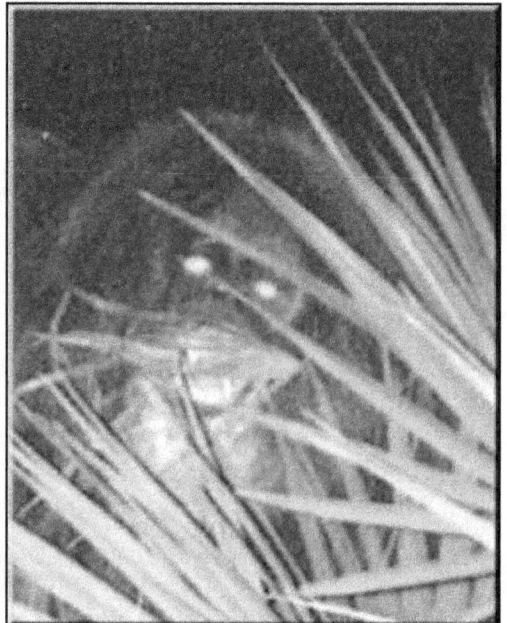

Your opinion is your opinion

Your Perception is your perception

Do not confuse them with facts or truth

Wars have been fought and millions have been killed

Because of inability of men to understand

The idea that EVERYBODY has a different view point

Chapter 14
Intelligence – Personality

Introduction

Among the millions of species that exist on the earth, the human being is said to be superior and exclusive (separate) because of its reasoning of distinguishing between right and wrong. The ability to adopt to the environment with and to master situations, understanding, ability to command and capacity to carry on difficult tasks by learning and putting the past experience to the most beneficent use. This quality, which we describe as intelligence is found in different degrees in different human beings.

Definitions

Intelligence is the ability of an individual to make profitable use of past experience –Thorndike Intelligence is the ability demanded in the solution of problems, which require the comprehension, and the use of symbols.

– Grprett

Intelligence is the ability of an individual to adjust himself to the conditions that arise in his environment.

– Brown

Intelligence is the ability to adopt oneself to judge well, understand well, reason (think) well and act well.

– Binet

Intelligence is the organization of abilities to learn a group of facts with alertness and accuracy to exercise mental control and display flexibility in seeking the solution of problem.

– Skinner

Types of Intelligence

According to Thorndike intelligence is of three types:

1. Abstract Intelligence or Cognitive Ability:

1. Abstract means which is not physically existing *e.g.* alphabets, numbers *etc.*
2. It is the ability to understand and deal with verbal and mathematical symbols
3. Of the three abilities abstract intelligence is one that receives greatest weight and almost pronounced as a correct test of intelligence
4. It is also the ability of manipulating ideas and relationships and more concerned with understanding abstract things
5. Philosophers and Professional people are high in abstract intelligence *e.g.*vocabulary, language, relational concepts *etc.*

Concrete Intelligence or Mechanical Intelligence or Motor Ability

1. Concrete means which is physically existing *e.g.* implement, object etc
2. It is the ability to understand and deal with things or objects *etc.*, and more concerned with the physical skills of individuals
3. Industrial and building traders are high in mechanical intelligence *e.g.* problem solving skill and manual skills

Social Intelligence or Social Ability

1. It is the ability to understand and deal with persons
2. It is the ability to understand and apply psychological principles of human relationships
3. Salesmen, politicians, leaders possess this intelligence *e.g.* association with people and empathy (understanding people by taking their conditions mentally)

An ideal person is one who has all the three types of intelligence.

Intelligence is the product of heredity and environment. Opportunities to learn vary widely, yet the inherited capacity (capacity taken by birth) as modified (changed or increased) by maturation (development) accounts for a greater part of the individual variability (differences in the intelligence of the individuals). The totality of biologically transmitted factors that influence the structure of body is referred as heredity.

Factors Affecting Intelligence

G.Brown a psychologist pointed out that, there are numerous factors which directly or indirectly affect the intelligence or abilities of the individual and which makeup the behaviour pattern of the individual.

Important Factors that Affect the Intelligence

1. **Heredity and environment**: heredity provides the physical body to be developed with certain inherent capabilities while environment provides maturation and training of the organism. Newman concludes that the variations in I.Q. or intelligence were determined about 68 per cent by heredity and 32 per cent by environment. It means that 68 per cent of intelligence of the individual comes through heredity and 32 per cent by environment

2. **Age:** The intelligence is maximum at 20 years and remains relatively stable if health and other factors do not interfere, until around 70 years when it rapidly decreases due to decline in physical efficiency

3. **Health and physical development:** Health and physical development are directly related to mental activity. Physical and physiological defects result in sub-normal intelligence or less intelligence

4. **Race:** As it is race has no influence over the intelligence but certain races which are socio-economically and culturally week show marginal effect on intelligence

5. **Sex:** Not much difference is noticed as per the sex of the individual. According to Crow and Crow males are slightly superior than females in questions that involve mathematical material and scientific concepts or in performance of certain scientific tasks (work related to science) and girls excel that deal more directly with the humanities (languages, literature, philosophy, fine arts, history *etc.*)

6. **Social and economic conditions**: if these conditions are good then physical development and mental development will also be fairly good and intelligence will be better

7. **Intelligence Quotient:** I.Q. rates the levels of intelligence of a person

$$I.Q. = \frac{\text{Mental age of an individual (MA)}}{\text{Chronological age of an Individual (CA)}} \times 100$$

I.Q. Level Character

Below 20 Idiot

20 to below 70 Feeble minded

70 to 90 Dull

90 to 110 Average or Normal

110 to 120 Superior

120 to 140 Very superior

140 to 200 Genius

Above 200 Supreme genius

Importance of Intelligence in Extension Work

1. Intelligence does not follow a set of stereo (similar) types of pattern but depends largely on the complexity of demand of their environment and the kind of training they receive

2. Intelligence remain constant when the conditions remain constant *i.e.* health, types of education and situation

3. In all, the differences in intelligence can be treated to either heredity or environment since individual is a product of both

4. Gifted persons with higher intelligence can be better utilized by offering broader opportunities and with programmes for their accelerated growth

5. It is easy to identify the mentally retarded people or people with less intelligence and problem men (persons with less intelligence due to physiological defects) in rural society and such people should be given special attention while training them in agricultural technologies

6. An extension worker can increase his effectiveness by using appropriate techniques for teaching farmers with different levels of intelligence and thereby smooth introduction of the programs of change

Concept of Personality

Personality is the total quality of an individual. The word personality comes from the Latin 'persona' which means the mask worn by players in the theater. Personality consists of observable behaviour. It is defined as an individual's typical or consistent adjustments to his environment.

The elements of personality are called the traits of personality; it is the traits that make one person different from another person in his behaviour. Shyness and sociability are different traits of personality. All characteristics which an individual possesses are his powers, needs, abilities, wants, habits, his goal and aspirations. His patterns of behaviour to objects and persons constitute his personality.

Since an individual is a bundle of characteristics as traits, we can define personality as an integrated pattern of traits.

According to Worth one's personality is made up of high physique, chemique, instrincts, and intelligence.

According to William James personality consists of ones

1. Material self (his body, clothes, family property *etc.*)

2. Social self (his home, club, office, church *etc.*)

3. Spiritual self (his ability to argue and discriminate, consciousness, moral sensitivities).

According to Medougali personality consists of his

a. Disposition

b Temperment and

c. Character

Definition of Personality

A men's personality is the total picture of his organised behaviour, especially as it can be characterized by his fellow men in a consistent way.

– Dashiell, IF

Our personality is the result of what we start with and what we have lived through. It is the' reaction mass' as a whole.

– Watson, J.B.

Personality is the dynamic organisation within the individual of those psychological systems that determine his unique adjustments to his environment

– Allport, G.W.

Personality Traits

Traits may be regarded as a dimension of personality. For example, dominance-submission is a trait that a person may show his any degree. A trait is a description of human behaviour. The traits of a person describe his personality. One's traits and the ways they are patterned make him different from other persons.

Openness to Experience (Inventive/curious vs. Consistent/cautious)

Appreciation for art, emotion, adventure, unusual ideas, curiosity, and variety of experience. Openness reflects the degree of intellectual curiosity, creativity and a preference for novelty and variety a person has. It is also described as the extent to which a person is imaginative or independent, and depicts a personal preference for a variety of activities over a strict routine. Some disagreement remains about how to interpret the openness factor, which is sometimes called "intellect" rather than openness to experience.

Sample Openness Otems

☆ I have excellent ideas., I am quick to understand things.

Conscientiousness (Efficient/organized vs. easy-going/careless)

A tendency to show self-discipline, act dutifully, and aim for achievement; planned rather than spontaneous behavior; organized, and dependable.

Sample Conscientiousness Items

☆ I am always prepared, I pay attention to details.

Extraversion (Outgoing/Energetic vs. Solitary/Reserved)

Energy, positive emotions, surgency, assertiveness, sociability and the tendency to seek stimulation in the company of others, and talkativeness.

Sample Extraversion Items

☆ I am the life of the party, I feel comfortable around people.

Agreeableness (Friendly/Compassionate vs. Cold/Unkind)

A tendency to be compassionate and cooperative rather than suspicious and antagonistic towards others. It is also a measure of one's trusting and helpful nature, and whether a person is generally well tempered or not.

Sample Agreeableness Items

☆ I feel others' emotions, I make people feel at ease.

Neuroticism (Sensitive/Nervous vs. Secure/Confident)

The tendency to experience unpleasant emotions easily, such as anger, anxiety, depression, or vulnerability. Neuroticism also refers to the degree of emotional stability and impulse control and is sometimes referred to by its low pole, "emotional stability".

Sample Neuroticism Items

☆ I am easily disturbed, I get irritated easily.

Personality Types

Jung has given this introversion - extroversion type in personality. Introverts will react negatively to situations, withdraw from the society, inwardly, selective. The introvert withdraw from the active participation in the objective world and he is interested in his inner world of thought and fantacy. He will not move freely with others. He won't express out his feelings and inner desires. He is sensitive to criticism, magnifies his failures and occupied with self criticism.

Extroverts are opposite to introverts. Extroverts react positively and outwardly expressive. An extrovert is supposed to be thick skinned and relatively sensitive to criticism, spontaneous in his emotional expression, impersonal in argument, neither deeply affected by failures nor much occupied with self analysis of self-criticism.

Sheldon's Classification (Types)

Sheldon classifies the people into 3 types:

1. Endomorphy
2. Mesomorphy
3. Ectomorphy

In addition to this sheldon also classified people into 3 types according to their temperaments. They are:

1. Viscerotonia
2. Somatotonia and
3. Cerebrotonia

Sl.No.	Body types	Temperamental types
1	Endomorphy	**Viscerotonia**
	Large Viscera	Love of comport affection
	Soft body contours	Sociability
2	Mesomorphy	**Somatotonia**
	Heavy muscular development Hard body contours	Vigorous, self -assertive
		Ambitious
3	Ecotomorphy	**Cerebrotonia**
	Long, slender extremities, poor	Restrained undiluted
	Muscular development	Social withdrawal thoughtful.

Measurement of Personality

The measurement of personality serves both the critical and practical purposes. When an individual who has difficulties of personnel adjustment comes for help to a psychiatrist, it is valuable to assess his personality. To measure the personality we have to measure the traits of personality. It is possible to measure needs, attitudes, interests, values and other personality characteristics also.

The following methods have been used for measuring personality characteristics.

The Questionnaire Method

This method involves the preparation of a list of questions or statements. Each one is concerned with some aspect of the feeling, attitude, habit or mode of behaivour related to the personality characteristic which is intended to be measured. The respondent is required to indicate his agreement or disagreement, acceptance or rejection, affirmation or denial of each statement. The questions are constructed in such a way that the affirmation of some and the denial of others express the presence of the trait that is measured. Each question is followed by 'Yes"No' 'Untrue' 'Doubtful' or other similar responses. Here the respondent under line a response that shows his character.

Example

1. In social gatherings, I like to be the centre of attention.
2. In a meeting, I prefer to occupy a back seat.
3. When you have to make an important decision, do you prefer some one else to decide for you.

Usually a personality test contains 30 or 50 items. This questionnaire method is also called objective test of personality. In this method the respondent should be truthful and very frank.

Projective Tests

Projective tests are so named because they induce the individual to project to put himself into the test situation or to identify with the person's therein and by telling about them, to reveal his own motives, attitudes, apprehensions and aspirations.

The use of projective test is based on the mechanism of projection. The projective test is a device for measuring the personality which the person tested does not recognise as such. He makes spontaneous responses in the test. The test presents a highly ambiquous situation which the tests is required to perceive and describe. The situation has no definite characteristics and so different persons perceive it in different ways. In perceiving the ambiquous situation, each person reflects his own needs, attitudes, habits, interests, feelings and behaviour patterns. The projective method is called projective because in perceiving and describing the situation, a person brings out his personality characteristics.

Two well known projective tests are:

1. Rorschach Inkblot Test
2. Thematic apperception Test.

1) Rorschach Test

Rorschach test was introduced by Hermann Rorschach. So the test was called after his name. This is the most widely used test. It was first described in 1921 by Hermann Rorschach, a Swiss psychiatrist.

Rorschach Test consists of 10 cards, each one having an inkblot on it. Some blots are coloured and some are in black and white. These cards are always presented in regular serial order. The inkblot does not represent any subject. It is ambiquous or unstructured figure. The ambiquity of the blot results in the great variety of responses it produces. Such an inkblot is placed before the subject. He is asked to describe what he sees. He is also required to say in what portion he sees and what makes him see the object that he describes. His responses are noted and classified under different categories.

These responses differ from person to person. Some respond to the whole figure, some to the parts of the figure. Some respond to form a shape, some to colour or the white space between the coloured or black patches. Some see moving and others stationery objectives. Some see human figures or parts, of human body. Others see animals. Classification of the responses made to the 10 cards by a subject shows that, certain types and categories are made consistently. These responses are interpreted as indicating the personality characteristics of the respondent.

2) Thematic Apperception Test (TAT)

This TAT is also a widely used projective Test. This TAT was first given by Murray. This test consists of 20 pictures. Each picture contains one or more persons in very ambiquous situations. For example, in one picture a very old woman is standing behind a young woman looking very serious. The subject is asked to write a story on each picture; the story should have a theme. It should say what is happening, and what is going to be the outcome or result.

In writing a story, the subject is expected to identify himself unconscientously with a character in the story. The story thus express out his own needs and frustrations, feelings and attitudes, ideas of self and of others, real or imaginary and so on. One cannot get clue about the characteristics of the subject from a single story. When one finds the recurrence of the same or similar characters, expressing similar feelings and attitudes, entertaining similar hopes and fears *etc.* one may be led to think that the stories reveal the person's own characteristics.

Personality Rating

In this method one person judges or rates the characteristic of another person. The person who judges is the rater and the person who is judged is the rate. The basis of the rating is the rater's general impression drawn from his observation of the behaviour of the rates in a variety of related situations. For example, a class teacher may be asked to judge such personality characteristics of his pupils as orderliness, punctuality, industriousness, co-operativeness *etc.* The teacher may be asked to express his rating on each trait by locating the position of each pupil on a so called *rating scale.*

Example of A Rating Scale *'Punctuality*

5	4	3	2	1
Always in	Generally	Some times	Generally	Never in time
punctual	punctual		unpunctual time	

If the teacher feels that the student is highly punctual he would place him at the left end of the scale and give him the score of 5.

Situational Test

In this situational test, a person is required to act in a situation which is specially arranged for the purpose of testing. The tester observes the behaviour of the testee while he is acting in the situation. Generally the testing is done in a group. For example, a party of 10 students is taken out and camps at the foot of hill. The students are given some cash. They are asked to plan and arrange for preparing their lunch. In this they will discuss and plan to report lunch. The teacher who accompanies them observes the behaviour of each and come to the conclusion regarding their personality character.

Interview

Interview is one of the methods to assess personality. The interview is a face to face situation consisting of the interviewee and the interviewer. The success of the interview depends upon the degree to which the interviewee makes free and frank responses. The interviewer should win the complete confidence of the interviewee.

He should establish with the interviewee of a relation of cordiality, warmth and responsiveness. Such a relation has been called *'rapport'* without a rapport no psychological interview can be a success.

Chapter 15
Teaching–Learning Process

Teaching Learning Process

Teaching

Teaching is the process of providing situations in which learning takes place; in other words, arranging situations in which the things to be learnt are brought to the attention of the learners, their interest is developed, desire aroused, conviction created, action promoted and satisfaction ensured.

Learning

Learning is a process by which a person becomes changed in his behaviour through self- activity.

– Leagans

"Learning is a process of progressive behaviour adaptation".

– Skinner

Process

Process means a course of procedures, something that occurs in a series of actions or events conducting to the desired end.

Learning Experience

It is the mental and/or physical reaction one makes through seeing, hearing or doing the things to be learned, through which one gains meanings and understandings of the material to be learned.

Learning is an active process on the part of the learner. Hence, a learning experience is not attained by mere physical presence in a learning situation. It is what the participant does (i. e. his reaction) while in the learning situation that is all-important in learning. He must give undivided attention to the instructor and deep thought to getting the facts, understanding their meaning, and to seeing their application to his needs and problems. Effective learning experi-ences, therefore, can best be had in effective learning situations provided by a skillful instructor who knows what he wants, who has the materials to accomplish his goals and the skills to use them effectively.

Learning Situation

A learning situation is a condition or environ-ment in which all the elements necessary for promoting learning are present; namely (1) Instructor (2) Learner (3) Subject matter (4) Teaching materials and equipment, and (5) Physical facilities.

An Effective Learning Situation

An effective learning situation is one in which all the essential elements for promoting learning *i.e.* learners, teachers, subject matter, teaching materials and physical facilities, relevant to a particular situation, are present in a dynamic relationship with one another. The conditions under which effective learning can take place are presented, following Leagans (1961).

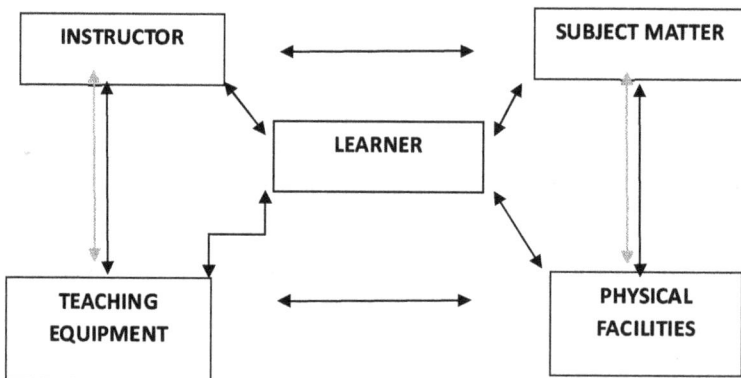

Elements of Learning Situation

1. Learners

Persons who want and need to learn are the learners. In an effective learning situation, learners occupy the most important central position and all efforts are directed towards them.

Learners should

    (i) be capable of learning,

    (ii) have interest in the subject,

(iii) have need for the information offered, and

(iv) be able to use the information once it is gained.

In the present context, the farmers, farm women and rural youth comprise the learners. To explain the learning situation, we take an example in which dairy farmers who need to increase milk production are the learners.

2. Teachers

They are the extension agents who impart training and motivate the learners. They not only know what to teach, but also know how to teach.

The teachers should

☆ have clear-cut and purposeful teaching objectives,

☆ know the subject matter and have it well organized,

☆ be enthusiastic and interested about the learners and the subject matter,

☆ be able to communicate and skillful in using teaching aids, and

☆ be able to encourage participation of the people.

3. Subject Matter

It is the content or topic of teaching that is useful to the learners.

☆ The subject matter should be

☆ pertinent to learners' needs,

☆ applicable to their real life situations,

☆ well organized and presented logically and clearly,

☆ consistent with the overall objectives, and

☆ challenging, satisfying and significant to the learners. Here, the subject matter is increasing milk production.

4. Teaching Materials

These are appropriate instructional materials, equipments and aids.

☆ The teaching materials should be-

☆ suitable to the subject matter and physical situation,

☆ adequate in quantity and available in time, and

☆ Skillfully used.

In the present example, the teaching materials may be improved breeds of bull or semen and fodder seeds suitable for the area, appropriate medicines, audio-visual aids relevant to the topic and level of understanding of the learners *etc.*

5. Physical Facilities

It means appropriate physical environment in which teaching-learning can take place.

☆ The physical facilities should be-

☆ compatible with the objective,

☆ representative of the area and situation, and

☆ adequate and easily accessible.

In the present example, physical facilities may include facilities for artificial insemination and administering medicines ; suitable land, irrigation *etc.* for growing fodder, and a place easily accessible, free from outside distractions, adequate seating arrangements, electricity for projection *etc.* for conducting training programme.

Criteria for Effective Learning

Learning involves acquisition of knowledge, skill, attitude *etc.* ; retention to stop reversion and transfer, to use it in real life situations. Learning to be effective, should have the following characteristics:

1. **Learning should he purposeful**: The learning must make sense and be useful to the learners. Objectives must be clear and meaningful to the learners. What is to be learnt must be important to, and wanted by a relatively large number of participants in the group, and must be attainable through the educational process, within the time limitations of the extension agent and the participants, within the physical and economic resources of the participants, and within the social condition and learning ability of the participants.

2. **Learning should involve appropriate activity by the learners that engages a maximum number of senses**: Messages reach the human mind through five senses, namely, seeing, hearing, feeling, tasting and smelling. In extension, most of the messages to be learnt reach the mind through seeing, hearing and doing. Learning should be experience centered *i.e.* farmers should primarily learn by doing, in addition to seeing and hearing.

3. **Learning must be challenging and satisfying**: Abilities acquired through learning should help the farmer to solve the problems, to overcome the difficulties and gradually lead to a more satisfying life.

4. **Learning must result in functional understanding**: Mere acquisition of knowledge is not enough; it must be understood and applied in real life situation.

Steps in Teaching (Extension Teaching)

Teaching is a planned and deliberate act on the part of the extension agent. The extension agent has to move step by step in a scientific and logical way to impart training to the clients who are farmers, farm women and rural youth. The role of the extension agent is that of a facilitator and motivator. Though details of the procedure may vary from situation to situation, there are some steps which are basic to extension teaching. These are presented following Wilson and Gallup (1955).

```
                              ┌─────────────────────────────┐
                              │ SATISFACTION                │
                        ┌─────┴─────────────────────────────┤
                        │ ACTION                            │
                   ┌────┴──────────────────────────────────┤
                   │ CONVICTION                             │
              ┌────┴───────────────────────────────────────┤
              │ DESIRE                                      │
         ┌────┴────────────────────────────────────────────┤
         │ INTEREST                                         │
    ┌────┴─────────────────────────────────────────────────┤
    │ ATTENTION                                             │
    └──────────────────────────────────────────────────────┘
```

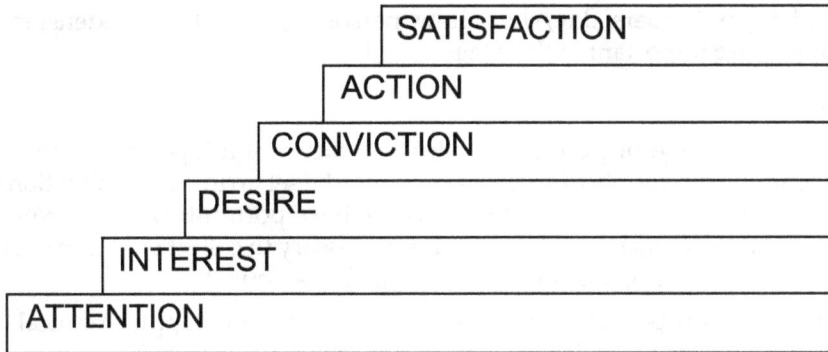

Steps in Extension Teaching

1. Attention

The first step in extension teaching is to make the people aware of new ideas and practices. The people must first know that a new idea, practice or object exists. This is the starting point for change. Until the individuals' attention have been focussed on the change that is considered desirable, there is no recognition of a problem to be solved or a want to be satisfied.

Mass methods like radio, television, exhibition, mobile phone *etc.* and personal contact by the extension agent, contact through local leaders are important at this stage.

2. Interest

Once the people have come to know of the new idea, the next step is to stimulate their interest. This may be done by furnishing them more information about the topic in a way they will be able to understand and use. It is necessary to present one idea at a time, relevant to their needs.

Personal contact by the extension agent, contact through local leaders, farm publications, radio, television *etc.* are important at this stage.

3. Desire

It means un-freezing the existing behaviour and motivating the people for change. At this stage it is necessary to emphasize on the advantages of the new idea or practice.

Visit to demonstrations, farm publications, personal contact by the extension agent, group discussion *etc.* are important at this stage.

4. Conviction

It is a stage of strong persuasion so as to convince the people about the applicability of the new idea or practice in their own situation and that it would be beneficial for them. The people are furnished with adequate information about the idea and how it works.

Field day or farmers' day, slide show, personal contact by the extension agent and training are important at this stage.

5. Action

This is the stage of putting the idea or practice into operation. Small scale demonstration with supply of critical inputs, may be set up in real life situation of the individuals who come forward. This Provides the opportunity of direct experience on the part of the learners. At this stage it is necessary to collect evidence of change such as change in yield, income, employment, behaviour *etc.*

Demonstration, personal contact by the extension agent, supply of critical inputs and ensuring essential services are important at this stage.

6. Satisfaction

To produce lasting change, the extension efforts should produce satisfying results. Satisfaction may come from high yield, more income, better health *etc.* Satisfaction reinforces learning and develops **dence**, which generates motivation for further change. To sustain the ged behaviour, it is necessary to furnish new and relevant information jut the practice on a continued basis, till change in the practice itself is **ieit** necessary.

Use of mass media, local leaders and personal contact by the extension agent are important at this stage. Availability of critical inputs and essential services are also to be ensured. The steps in extension teaching are to be synchronized with the Innovation-decision process of the adopters.

Principles of Learning as Applicable to Extension

There are some principles of learning which are very well applicable in extension. The principles may provide good guidance for making learning in extension effective.

These are:

1. *Principle of self-activity*: Learning is an active process on the part of the learners. The kind of learning which takes place is the result of the kind of experience one acquires. The experiences must be compelling and direct on the part of the learners.

   Conducting demonstration by the farmers in their own fields provides opportunity of self-activity, *i.e.* learning by doing. This makes learning effective and permanent.

2. *Principle of association*: New learning may be associated with previous successful and satisfying responses.

   If the farmers have obtained profitable return by the application of nitrogenous fertilizer, they may be motivated to use balanced fertilizer containing phosphate and potash, for still higher return.

3. *Principle of transfer*: Application of perceived relationship to another situation in which it is applicable. Unless knowledge or learning can be applied in a new situation, it remains very much restricted.

If the farmers have learnt the technique of water management in a particular crop, they should be able to use this knowledge in other crops as *mcU*. This shall spread the effect of learning.

4. *Principle of disassociation*: For effective learning, undesirable responses are to be eliminated. This may be done by setting up desirable substitutes which are more satisfying. When planting a crop in lines gives better yield, the farmers may be advised not to practise broadcasting.

5. *Principle of readiness*: Learning takes place more effectively when **one** is ready to learn. When farmers are ready to cooperate, with good guidance, they may **be** able to form a cooperative society.

6. *Principle of set or attitude*: An unfavourable attitude or set retards learning and a favourable attitude accelerates it. Unless attitude becomes favourable, adoption will not take place.

When farmers develop a favourable attitude towards scientific treatment of cattle, they shall learn the importance of this type of treatment **for** animals.

7. *Principle of practise*: Perfection is seldom achieved without practise. The practise must be correct, otherwise there will be wrong learning. The attainment of perfection demands that undesirable and useless movements are replaced by desirable and useful ones. Learning to use a sprayer correctly, requires practise several times over.

8. *Principle of motivation*: Motivation or drive means stimulation towards action. Without motivation an organism does not behave and hence does not learn. The practice recommended must be motivating for learning to take place. Favourable experience of planting trees motivate tribal farmers to collect saplings from the forest nursery.

9. *Principle of timing*: Other things being equal, learning takes place more readily when there is introduction of a topic or skill at a time when it can be used in some serviceable manner. When insects have appeared or are likely to appear in crops, farmers shall readily learn about plant protection.

10. *Principle of clarity of objectives*: The objective of learning should be clear. The ease of learning seems to vary directly with the meaningfulness of the material presented. Meaningful learning is interesting and easier than senseless learning.

When farmers use crop loan only for growing crops, they are clear about the objective of getting the loan. This clear understanding enable the farmers to learn about proper utilization and repayment of loans and take further loans if necessary, for economic development.

11. *Principle of satisfyingness*: A satisfying after-effect reinforces learning.

Crops grown during the rabi-summer season give higher economic return and higher level of satisfaction to the farmers. Farmers learn to invest more and take more care for crops during the season.

Principles of Teaching

1. Share intellectual control with students.

Building a sense of shared ownership is an effective way of achieving high levels of student interest and engagement. It can be achieved in many ways; many of these involve some form of formal or informal negotiation about parts or all of the content, tasks or assessment. Another complementary approach is to ensure that students' questions, comments and suggestions regularly influence, initiate (or terminate) what is done.

2. Look for occasions when students can work out part (or all) of the content or instructions.

Learning is almost always better if students work something out for themselves, rather than reading it or hearing it. This is not always feasible of course, but often it is. It can involve short, closed tasks: *e.g.* 'if the units of density are grams per cm work out the formula by which we calculate the density of a substance from the volume and mass of an object made of that substance'. It can also involve much longer open-ended tasks: *e.g.*'Here is a photo of the ruins of Machu Pichu, work out as much as you can, from this photo, about the Incas and their fate'.

3. Provide opportunities for choice and independent decision-making.

Students respond very positively to the freedom to make some decisions about what or how they will work. To be effective, the choices need to be genuine, not situations where there is really only one possibility. These may include choices about which area of content to explore, the level of demand (do more routine tasks or fewer more demanding ones), the form of presentation (poster, powerpoint presentation, role play, model *etc.*),and how to manage their time during a day or lesson.

4. Provide diverse range of ways of experiencing success.

Raising intellectual self-esteem is perhaps the most important aspect of working with low and moderately achieving students. Success via interactive discussion, question-asking, role-plays and tasks allowing high levelsof creativity often results in greater confidence and hence persistence in tackling other written tasks. Publicly recognising and praising good learning behaviours is useful here.

5. Promote talk which is exploratory, tentative and hypothetical.

This sort of talk fosters link-making and, as our research shows, commonly reflects high levels of intellectual engagement. Teaching approaches such as delayed judgement, increased wait-time, promotion of 'What If' questions and use of P.O.Es are all helpful. The classroom becomes more fluid and interactive.

6. Encourage students to learn from other students' questions and comments.

The (student) conception that they can learn from other students ideas, comments and questions develops more slowly than the conception that discussion is real and useful work. The classroom dynamics can reach new, very high levels when ideas and debate bounce around from student to student, rather than student to teacher.

7. Build a classroom environment that supports risk-taking.

We underestimated the very high levels of perceived risk that accompanies many aspects of quality learning for most students, even in classes where such learning is widespread. It is much safer, for example, to wait for the teacher's answer to appear than to suggest one yourself. Building trusts in the teacher and other students and training students to disagree without personal put-downs are essential to widespread display of good learning behaviours.

8. Use a wide variety of intellectually challenging teaching procedures.

There are at least two reasons for this, one is that teaching procedures that counter passive learning and promote quality learning require student energy and effort. Hence they need to be varied frequently to retain their freshness. The other is that variety is another source of student interest.

9. Use teaching procedures that are designed to promote specific aspects of quality learning.

One of the origins of PEEL was the belief that students could be taught how to learn, in part by devising a range of teaching procedures to variously tackle each of a list of poor learning tendencies, for example failing to link school work to relevant out-of-school experiences. The variety in (8) is not random and one basis for selecting a particular teaching procedure is to promote a particular aspect of quality learning.

10. Develop students' awareness of the big picture: how the various activities fit together and link to the big ideas.

Many, if not most students, do not perceive schooling to be related to learning key ideas and skills. Rather, they see their role as completing tasks and so they focus on what to do not why they are doing it. Much teacher talk, particularly in skills based areas such as Mathematics, Grammar and Technology reinforces this perception. For these reasons, students (including primary students) commonly do not link activities and do not make links to unifying, 'big ideas'.

11. Regularly raise students' awareness of the nature of different aspects of quality learning.

This is a key aspect of learning how to learn. Students typically have no vocabulary to discuss learning. it is very helpful to build a shared vocabulary and shared understandings by regular, short debriefing about some aspect of the learning that has just occurred. Having a rotating student monitor of a short list of good learning behaviours can be very helpful.

12. Promote assessment as part of the learning process.

Students (and sometimes teachers) typically see assessments as purely summative: something that teachers do to students at the end of a topic.Building the perception that (most) assessment tasks are part of the learning process includes encouraging students learning from what they did and did not do well as well as having students taking some ownership of and responsibility for aspects of assessment. It also includes teachers ensuring that they are assessing for a range of aspects of quality learning (*e.g.* if you want students linking different lessons then reward that in your assessment) and for a wider range of skills than is often the case.

Types of Learning

1. Perceptual Learning

Ability to learn to recognize stimuli that have been seen before:

☆ Primary function is to identify and categorize objects and situations

☆ Changes within the sensory systems of the brain

2. Stimulus-response Learning

Ability to learn to perform a particular behavior when a certain stimulus is present:

☆ Establishment of connections between sensory systems and motor systems

☆ **Classical conditioning** – association between two stimuli

☐ **Unconditioned Stimulus (US), Unconditioned Response (UR), Conditioned Stimulus (CS), Conditioned Response (CR)**

☐ **Hebb rule** – if a synapse repeatedly becomes active at about the same time that the postsynaptic neuron fires, changes will take place in the structure or chemistry of the synapse that will strengthen it

☐ Rabbit experiment – tone paired with puff of air

☐ **Instrumental conditioning** – association between a response and a stimulus; allows an organism to adjust its behavior according to the consequences of that behavior

☐ **Reinforcement** – positive and negative

☐ **Punishment**

3. Motor Learning

Establishment of changes within the motor system

4. Relational Learning

Involves connections between different areas of the association cortex

5. Spatial Learning

Involves learning about the relations among many stimuli

6. Episodic Learning

Remembering sequences of events that we witness

7. Observational Learning

Learning by watching and imitation other people

Perceptual Learning

☆ Involves learning **about** things, not what to do when they are present

☆ Simple perceptual learning, recognizing stimuli, takes place in appropriate regions of sensory association cortex

1. Visual Learning

☆ **Inferior temporal cortex** - necessary for visual pattern discrimination, receives info from visual cortex:

☆ **Ventral/dorsal streams** - what and where

☆ **Delayed matching-to-sample task** - requires that stimulus be remembered for a period of time:

☐ "remembering" the stimulus involves a neuronal circuit; it is the circuits, not the individual neurons that recognize particular stimuli

☐ lesions of inferior temporal cortex disrupts an animal's ability to remember what it has just seen

☐ electrical stimulation of the inferior temporal cortex during the delay causes forgetting

☆ Responses of single neurons in inferior temporal cortex recorded when pairs of stimuli shown:

☐ Found that when stimuli are paired, the neural circuits responsible for recognizing them become linked together.

☐ Perception of either stimulus activates both circuits.

☆ Visual long-term memory involves the establishment of new circuits in the inferior temporal cortex by means of synaptic changes.

2. Auditory Learning

☆ Auditory learning tasks modify response characteristics of neurons in various parts of the auditory system

☐ Study pairing tones with shock

▲ Pretraining, neuron responds best to 9.5 Hz tone

▲ CS is 9-Hz tone, paired with shock

▲ After training, neuron now responds best to 9-Hz tone, and less to 9.5 Hz

☆ **Medial division of the medial geniculate nucleus (MGm)** - important for classically conditioned emotional responses.

☐ Receives info from auditory and somatosensory systems

☐ Directly connected to the central nucleus of the amygdala

☐ Which activates the nucleus basalis

☐ Nucleus basalis contains acetylcholine neurons which innervate auditory cortex, telling it to pay particular attention to the ventral division of the medial geniculate nucleus (conveys auditory information).

Stimulus-response Learning

1. Classical Conditioning

☆ Central nucleus of the amygdala involved in classically-conditioned emotional responses

☆ MGm eceives auditory and somatosensory info

☆ Pairing of tone and footshock increases responses to CS

☆ **Extinction** - somehow, NMDA receptors again involved - most likely due to long-term depression.

2) Instrumental Conditioning and Motor Learning

☆ 2 pathways exist between sensory association cortex and motor association cortex:

    1) **Direct trascortical connections** - short-term memory, and with hippocampus involved in episodic memory

    2) **Via basal ganglia and thalamus** - used once no longer "new" learning; Parkinson's example

☆ **Premotor cortex** - monkeys raising arms for food; humans learning motor task

☆ **Reinforcement:**

☆ Reinforcing brain stimulation discovered by Olds and Milner in 1954

☆ **Medial forebrain bundle** - best place for self-stimulation; also involved in natural reinforcers, such as food, water, or sex

☆ **Dopamine** - likely serves as a neuromodulator, involved in reinforcement - receptors in **nucleus accumbens**

    ❐ **Conditioned place preference** - animals prefer to be where they have encountered a reinforcing stimulus

    ❐ This process used to test influence of dopamine on reinforcement

    ❐ If given dopamine antagonist during place training, don't develop conditioned place preference

    ❐ Also involved in **negative reinforcement** - dopamine antagonists prevent avoidance learning

    ❐ Amphetamine is a dopamine agonist - animals will work to get injections of it

        ▲ **drug discrimination procedure** - train animals to press a certain lever to receive food after it has been given a drug and to press another lever after it has been given saline

        ▲ found that rats would press drug lever only if amphetamine put directly into nucleus accumbens, rather than other brain structure

▲ they pushed the saline lever if they also had been given a dopamine receptor blocker with the amphetamine into the nucleus accumbens.

Reinforcement System

☆ Must be able to first detect and then respond to reinforcement opportunities

☆ **Ventral tegmentum area** - reinforcing stimuli activate neurons here, which they stimulate release of dopamine in other systems

　　☐ neurons activated by both natural and conditioned reinforcers

　　☐ information received from 3 sites:

　　　　1) **Amygdala** - involved in detection of CS for reinforcement - if monkeys trained that food follows a visual stimulus, then amygdala lesioned, the monkey forgets the association

　　　　2) **Lateral hypothalamus** - neurons become active when monkeys see food, but only when hungry - neurons show sensory-specific satiety; activity related to presence of reinforcing stimuli

　　　　3) **Prefrontal cortex** - secretes excitatory glutamate, which triggers bursts of dopamine to be released from neurons in the ventral tegmental area into the nucleus accumbens; may serve as monitor for reinforcement-seeking activity

☆ A reinforcement system is required by instrumental conditioning

　　1) Discriminative stimulus activates weak synapse

　　2) Circumstance that causes animal to press lever activates a strong synapse

　　3) If behavior is reinforced, then neurotransmitter/neuromodulator released (dopamine), causing synaptic changes, strengthening weak synapses

Everyone processes and learns new information in different ways. There are three main cognitive learning styles: visual, auditory, and kinesthetic. The common characteristics of each learning style listed below can help you understand how you learn and what methods of learning best fits you. Understanding how you learn can help maximize time you spend studying by incorporating different techniques to custom fit various subjects, concepts, and learning objectives. Each preferred learning style has methods that fit the different ways an individual may learn best.

Common Characteristics

Visual

☆ Uses visual objects such as graphs, charts, pictures, and seeing information

☆ Can read body language well and has a good perception of aesthetics

☆ Able to memorize and recall various information

☆ Tends to remember things that are written down

☆ Learns better in lectures by watching them

Auditory

☆ Retains information through hearing and speaking

☆ Often prefers to be told how to do things and then summarizes the main points out loud to help with memorization

☆ Notices different aspects of speaking

☆ Often has talents in music and may concentrate better with soft music playing in the background

Kinesthetic

☆ Likes to use the hands-on approach to learn new material

☆ Is generally good in math and science

☆ Would rather demonstrate how to do something rather than verbally explain it

☆ Usually prefers group work more than others.

☆ **Visual:** The occipital lobes at the back of the brain manage the visual sense. Both the occipital and parietal lobes manage spatial orientation.

☆ **Aural:** The temporal lobes handle aural content. The right temporal lobe is especially important for music.

☆ **Verbal:** The temporal and frontal lobes, especially two specialized areas called Brocaï¿½s and Wernickeï¿½s areas (in the left hemisphere of these two lobes).

☆ **Physical:** The cerebellum and the motor cortex (at the back of the frontal lobe) handle much of our physical movement.

☆ **Logical:** The parietal lobes, especially the left side, drive our logical thinking.

☆ **Social:** The frontal and temporal lobes handle much of our social activities. The limbic system (not shown apart from the hippocampus) also influences both the social and solitary styles. The limbic system has a lot to do with emotions, moods and aggression.

☆ **Solitary:** The frontal and parietal lobes, and the limbic system, are also active with this style.

Chapter 16

Motivation

Basic Needs

A motivating force that compels action for its satisfaction. Needs range from basic survival needs (common to all human beings) satisfied by necessities, to cultural, intellectual, and social needs (varying from place to place and age group to age group) satisfied by necessaries. Needs are finite but, in contrast, wants are boundless.

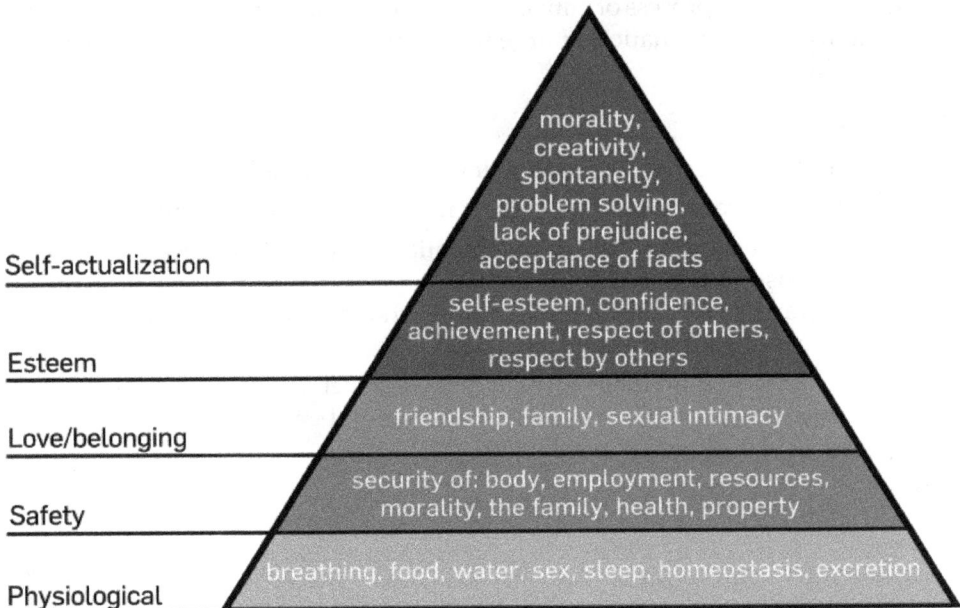

Self-actualization	morality, creativity, spontaneity, problem solving, lack of prejudice, acceptance of facts
Esteem	self-esteem, confidence, achievement, respect of others, respect by others
Love/belonging	friendship, family, sexual intimacy
Safety	security of: body, employment, resources, morality, the family, health, property
Physiological	breathing, food, water, sex, sleep, homeostasis, excretion

Maslow's hierarchy of needs which focus on describing the stages of growth in humans. Maslow used the terms Physiological, Safety, Belongingness and Love, Esteem, Self-Actualization and Self-Transcendence needs to describe the pattern that human motivations generally move through.

Classification of Needs

The desire for security:

Economic, social, psychological and spiritual security. Man wants protection for his physical being - food, clothing and shelter. It may also mean an adequate reserve of wealth to secure more material things in the future. The wish for security may also be satisfied by spiritual beliefs. In fact, in history whole cultures have put emphasis on security. The Great Wall China, the Maginet Line, the Tower of Babel, the innumerable forts and fortresses in several countries are striking examples

The desire for affection or response: companionship gregariousness, and social mindedness, the need for a feeling of belonging.

The desire for recognition*:* status, prestige, achievement and being looked upto. Each individual feels the need to be considered important by his fellowmen.

The desire for new experience : adventure, new interests, new ideas, new friends and new ways of doing things. Some people primarily want the thrill of something new, something different.

Organic needs: Organic needs like sex, hunger and thirstyness are also very important for human beings. The above five categories represent all the powerful motivating forces stated in general.

Motivation

Motivation is the process of initiating a conscious and purposeful action. Motive means an urge, or combination of urge to induce conscious or purposeful action. It is goal -directed.

Definition

(i) The goal directed, need satisfying behaviour is called motivation.

(ii) It is a process of initiating a conscious and purposeful action.

(iii) Motive means an urge or combination of urges to induce conscious or purposeful action. Motives, arising out of natural urges or acquired interests, or dynamic forces that affect thoughts, emotions and behaviour, *eg.:* Motive for a murder

Motivation is an inner drive to behave or act in a certain manner. These inner conditions such as wishes, desires and goals, activate to move in a particular direction in behavior.

Types

Intrinsic and Extrinsic Motivation

Motivation can be divided into two types: intrinsic (internal) motivation and extrinsic (external) motivation.

Intrinsic Motivation

Intrinsic motivation refers to motivation that is driven by an interest or enjoyment in the task itself, and exists within the individual rather than relying on external pressures or a desire for reward. Intrinsic motivation is a natural motivational tendency and is a critical element in cognitive, social, and physical development. Students who are intrinsically motivated are more likely to engage in the task willingly as well as work to improve their skills, which will increase their capabilities. Students are likely to be intrinsically motivated if they:

☆ attribute their educational results to factors under their own control

☆ believe they have the skills to be effective agents in reaching their desired goals, also known as self-efficacy beliefs

☆ are interested in mastering a topic, not just in achieving good grades

Extrinsic Motivation

Extrinsic motivation refers to the performance of an activity in order to attain an outcome, whether or not that activity is also intrinsically motivated. Extrinsic motivation comes from outside of the individual. Common extrinsic motivations are rewards (for example money or grades) for showing the desired behavior and the threat of punishment following misbehavior. Competition is in an extrinsic motivator because it encourages the performer to win and to beat others, not simply to enjoy the intrinsic rewards of the activity. A cheering crowd and the desire to win a trophy are also extrinsic incentives.

Functions of Motivation

(i) Motives encourage a learner in his learning activities.

(*e.g.*) Extrinsic motives like prizes, medals *etc.* motivate

(ii) Motives act as selectors of the type of activity in which the person desires to engage.

(*e.g.*) Selection of courses,

(iii) Motives direct and regulate behaviour.

(*e.g.*) discipline in schools *etc.*

Techniques of Motivation

(1) Need Based Approach

The approach should be need based so that it could satisfy five categories of need by knowing the level of motivation and patterns of motivation among them.

The five categories of needs are (i) physiological need (ii) desire for security (iii) desire for recognition (iv) desire for new experiences and (v) organic needs.

(2) Training to Set a Realistic Level of Aspirations

Any attempt to revise the expectations of farmer's should be done with full understanding of their socio-economic status (*e.g.*):

(i) Creating an aspiration in a farmer who doesn't have any land of his own for possession of one or two acres.

(ii) A person who attains 30 tonnes/acres of yield could be made to aspire for 40 tonnes/acre. Such a realistic levle of aspiration would ensure slow and steady progress.

(3) Participation

The involvement of farmers in the programmes of agricultural change acts as booster of motivation not only for the immediate participants but also for others.

(4) Use of Audio Visuals

The proper selection, combination and use of various audio visuals for the appropriate purpose will act as lubricants of motivation.

Importance of Motivation in Extension

1. For mobilising the villagers and extension workers.

2. Knowledge of biological drive/need helps the extension worker to realise the problems of the people. It helps in sympathetic handling.

3. Knowledge of psychological and social drives helps the extension worker to formulate progrmmes and make effective approaches in changing their attitude.

Knowledge of other motivating forces helps avoiding conflicts or tensions.

Chapter 17
Attitude

An **attitude** is an expression of favor or disfavor toward a person, place, thing, or event.

Attitudes involve some knowledge of a situation. However, the essential aspect of the attitude is found in the fact that some characteristic feeling or emotion is experienced and, as we would accordingly expect, some definite tendency to action is associated. Subjectively, then, the important factor is the feeling or emotion. Objectively it is the response, or at least the tendency to respond. Attitudes are important determinants of behaviour. If we are to change them we must change the emotional components.

Allport has defined attitude as a mental and neutral state of readiness organised through experience, exerting a directive or dynamic influence upon the individual's response to all objects with which it is related.

A farmer may vote for a particular political party because he has been brought up to believe that it is the "right" party. In the course of experience he may learn some thing about the policies of that party. In that case his attitude will probably change. As a result, he may be expected to vote in a different way. Knowledge, attitudes and behaviour are then very closely linked.

Attitude Change

Well established attitudes tend to be resistant to change, but others may be more amenable to change. Attitudes can be changed by a variety of ways. Some of the ways of attitude change are as follows:

1. By obtaining new information from other people and mass media, resulting in changes in cognitive component of a person's attitudes.

2. Attitudes may change through direct experience.

3. Attitudes may change through legislation.

4. Since person's attitudes are anchored in his membership group and reference groups, one way to change the attitude is to modify one or the other.

5. Attitude change differs with reference to the situation also.

Factors Influencing the Development of Attitudes

1. Maturation

The young **child** has **only** a **very limited capacity for** understanding the world-about him and he is consequently incapable of forming attitudes about remote, or complex, or abstract things or problems.

At about a mental age of twelve years **the child** begins to understand abstract terms **such** as **pity and justice,** and his capacity for both inductive and deductive reasoning shows a marked and continuous increase during adolescence. As a result of this growth in capacity, he becomes able to understand and react to more abstract and more generalized propositions, ideas and ideals.

At the age of four or five years, three characteristics especially deserve mention. These are curiosity, central-suggestibility, and independence. The child at this age is likely to express his curiosity by asking an endless series of questions.

Adolescence is marked especially by the maturation of sex emotions and by the development of altruism and co-operativenes. These in large measure furnish the basis for the formation of attitudes that differentiate adults from children.

2. Physical Factors

Clinical psychologists have generally recognized that physical health and vitality are important factors in determining adjustment and frequently it has been found that malnutrition or disease or accidents have interfered so seriously with normal development that serious behaviour disturbances have followed.

3. Home Influences

It is generally accepted that attitudes are determined largely by social environment and that home influences are especially important.

4. The Social Environment

The home environment is of primary importance in the formation of early attitudes, but friends, associates, and the general social environment come to have an increasing influence as the child grows older and has wider social contacts.

5. Government

The form of the government seems to be an important factor in determining attitudes both towards government itself and towards other things.

6 Movie Pictures

Attendance at movie pictures constitutes another important possible influence in determination of attitudes. Thurstone concluded that films definitely changed the social attitudes.

7. The Teacher

Teacher may be one of the influential factors for attitude change. Brown concluded that the personalities of their teachers had been the most important single factor for attitude change.

8. The Curriculum

Curriculum of the course also played very important role in attitude change. Thorndike concluded from his research that curriculum of the course played major share in attitude change among students.

9. Teaching Methods

One of the categories in Brown's study was *"manner of presentation"* of subject matter. This was judged to have a favourable effect by 80.0 per cent of the students and an unfavfourable effect by 17.7 per cent.

Chapter 18

Data Collection and Methods

Data collection is a term used to describe a process of preparing and collecting data, for example, as part of a process improvement or similar project. The purpose of data collection is to obtain information to keep on record, to make decisions about important issues, or to pass information on to others. Data are primarily collected to provide information regarding a specific topic.[1]

The **primary data** are those which are collected afresh and for the first time, and thus happen to be original in character. Such data are published by authorities who themselves are responsible for their collection.

The **secondary data,** on the other hand, are those which have already been collected by some other and which have alread been processed. Generally speaking, secondary data are information which have previously been collected by some organisation to satisfy its own need but it is being used by the department under reference for an entirely different reason. For example, the census figures are published every tenth year by the Registrar General of India. But the census figures are also used by demographers and other social scientists for planning and research. Thus, the officials of the department of Registrar General will visualise the census figures as primary data. But a demographer using the same census figures to prepare a mortality table will consider them as secondary data.

Data Collection is an important aspect of any type of research study. Inaccurate data collection can impact the results of a study and ultimately lead to invalid results.

Data collection methods for impact evaluation vary along a continuum. At the one end of this continuum are quantatative methods and at the other end of the continuum are Qualitative methods for data collection.

Observation Method

The investigator collects the requisite information personally through observation. For example, in order to study the conditions of students residing in a university, the investigator meets the students in their hostels and collects necessary data after a personal study. The information about the extent of damage caused by natural calamities like flood can be collected by personal observation by a trained investigator. As the investigator is solely responsible for Collection of data by this method, his training, skill and knowledge play an important role on the quality of primary data.

A slight variation of this procedure is indirect **oral investigation** where data are collected through indirect sources. Persons who are likely to have information about the problem are interrogated and on the basis of their answers, primary data become available. Most of the Commissions of Enquiry or Committees appointed by Government collect primary data by this method. The accuracy of the primary data collected by this method depends largely upon the type of persons interviewed and hence these persons have to be selected very carefully.

Questionnaire Method

A popular and common method of collection of primary data is by personally interviewing individuals, recording their answers in a structured questionnaire. The complete enumeration of Indian decennial census is performed by this method. The enumerators visit the dwellings of individuals and put questions to them which elicit the relevant information about the subject of enquiry. This information is recorded in the questionnaire. Occasionally a part of the questionnaire is unstructured so that the interviewee can feel free to share information about intimate matters with the interviewer. As the data are collected by the field staff personally it is also known as personal interview method.

Much of the accuracy of the collected data, however, depends on the ability and tactfulness of investigators, who should be subjected to special training as to how they should elicit the correct information through friendly discussions.

Mailed Questionnaire Method

A set of questions relevant to subject of enquiry are mailed to a selected list of persons with a request to return them duly filled in. Supplementary instructions regarding the definitions of terms used and the methods of filling up the forms should also accompany the questionnaire. This method can only be used when the respondents are literate and can answer the questions in writing. The questions should be very clear without any ambiguity keeping in mind that there is no investigator to help the respondent.

The method of collecting data by mailing the questionnaires to the respondents is most extensively employed in various business and economic surveys. This method saves both time and cost and can cover a large area. The absence of an investigator, however, renders the responses less reliable. The method also suffers from a large degree of non response.

In primary data collection, you collect the data yourself using methods such as interviews and questionnaires. The key point here is that the data you collect is unique to you and your research and, until you publish, no one else has access to it.

There are many methods of collecting primary data and the main methods include:

☆ Questionnaires

☆ Interviews

☆ Focus group interviews

☆ Observation

☆ Case-studies

☆ Diaries

☆ Critical incidents

☆ Portfolios.

The primary data, which is generated by the above methods, may be qualitative in nature (usually in the form of words) or quantitative (usually in the form of numbers or where you can make counts of words used). We briefly outline these methods but you should also read around the various methods. A list of suggested research methodology texts is given in your *Module Study Guide* but many texts on social or educational research may also be useful and you can find them in your library.

Questionnaires

Questionnaires are a popular means of collecting data, but are difficult to design and often require many rewrites before an acceptable questionnaire is produced.

Advantages

☆ Can be used as a method in its own right or as a basis for interviewing or a telephone survey.

☆ Can be posted, e-mailed or faxed.

☆ Can cover a large number of people or organisations.

☆ Wide geographic coverage.

☆ Relatively cheap.

☆ No prior arrangements are needed.

☆ Avoids embarrassment on the part of the respondent.

☆ Respondent can consider responses.

☆ Possible anonymity of respondent.

☆ No interviewer bias.

Disadvantages

☆ Design problems.

☆ Questions have to be relatively simple.

☆ Historically low response rate (although inducements may help).

☆ Time delay whilst waiting for responses to be returned.

☆ Require a return deadline.

☆ Several reminders may be required.

☆ Assumes no literacy problems.

☆ No control over who completes it.

☆ Not possible to give assistance if required.

☆ Problems with incomplete questionnaires.

☆ Replies not spontaneous and independent of each other.

☆ Respondent can read all questions beforehand and then decide whether to complete or not. For example, perhaps because it is too long, too complex, uninteresting, or too personal.

Interviews

Interviewing is a technique that is primarily used to gain an understanding of the underlying reasons and motivations for people's attitudes, preferences or behaviour. Interviews can be undertaken on a personal one-to-one basis or in a group. They can be conducted at work, at home, in the street or in a shopping centre, or some other agreed location.

Advantages

☆ Serious approach by respondent resulting in accurate information.

☆ Good response rate.

☆ Completed and immediate.

☆ Possible in-depth questions.

☆ Interviewer in control and can give help if there is a problem.

☆ Can investigate motives and feelings.

☆ Can use recording equipment.

☆ Characteristics of respondent assessed – tone of voice, facial expression, hesitation, *etc.*

☆ Can use props.

☆ If one interviewer used, uniformity of approach.

☆ Used to pilot other methods.

Disadvantages

☆ Need to set up interviews.

☆ Time consuming.

☆ Geographic limitations.

☆ Can be expensive.

☆ Normally need a set of questions.

☆ Respondent bias – tendency to please or impress, create false personal image, or end interview quickly.

☆ Embarrassment possible if personal questions.

☆ Transcription and analysis can present problems – subjectivity.

☆ If many interviewers, training required.

Types of Interview

Structured

☆ Based on a carefully worded interview schedule.

☆ Frequently require short answers with the answers being ticked off.

☆ Useful when there are a lot of questions which are not particularly contentious or thought provoking.

☆ Respondent may become irritated by having to give over-simplified answers.

Semi-structured

The interview is focused by asking certain questions but with scope for the respondent to express him or herself at length.

Unstructured

This also called an in-depth interview. The interviewer begins by asking a general question. The interviewer then encourages the respondent to talk freely. The interviewer uses an unstructured format, the subsequent direction of the interview being determined by the respondent's initial reply. The interviewer then probes for elaboration – 'Why do you say that?' or, 'That's interesting, tell me more' or, 'Would you like to add anything else?' being typical probes.

The following section is a step-by-step guide to conducting an interview. You should remember that all situations are different and therefore you may need refinements to the approach.

Planning an Interview

☆ List the areas in which you require information.

☆ Decide on type of interview.

☆ Transform areas into actual questions.

☆ Try them out on a friend or relative.

☆ Make an appointment with respondent(s) – discussing details of why and how long.

☆ Try and fix a venue and time when you will not be disturbed.

Conducting an Interview

☆ Personally: arrive on time be smart smile employ good manners find a balance between friendliness and objectivity.

☆ At the start: introduce yourself re-confirm the purpose assure confidentiality – if relevant specify what will happen to the data.

☆ The questions: speak slowly in a soft, yet audible tone of voice control your body language know the questions and topic ask all the questions.

☆ Responses: recorded as you go on questionnaire written verbatim, but slow and time-consuming summarised by you taped – agree beforehand – have alternative method if not acceptable consider effect on respondent's answers proper equipment in good working order sufficient tapes and batteries minimum of background noise.

☆ At the end: ask if the respondent would like to give further details about anything or any questions about the research thank them.

Telephone Interview

This is an alternative form of interview to the personal, face-to-face interview.

Advantages

☆ Relatively cheap.

☆ Quick.

☆ Can cover reasonably large numbers of people or organisations.

☆ Wide geographic coverage.

☆ High response rate – keep going till the required number.

☆ No waiting.

☆ Spontaneous response.

☆ Help can be given to the respondent.

☆ Can tape answers.

Disadvantages

☆ Often connected with selling.

☆ Questionnaire required.

☆ Not everyone has a telephone.

☆ Repeat calls are inevitable – average 2.5 calls to get someone.

☆ Time is wasted.

☆ Straightforward questions are required.

☆ Respondent has little time to think.

☆ Cannot use visual aids.

☆ Can cause irritation.

☆ Good telephone manner is required.

☆ Question of authority.

Focus Group Interviews

A focus group is an interview conducted by a trained moderator in a non-structured and natural manner with a small group of respondents. The moderator leads the discussion. The main purpose of focus groups is to gain insights by listening to a group of people from the appropriate target market talk about specific issues of interest.

Observation

Observation involves recording the behavioural patterns of people, objects and events in a systematic manner. Observational methods may be:

☆ Structured or unstructured

☆ Disguised or undisguised

☆ Natural or contrived

☆ Personal

☆ Mechanical

☆ Non-participant

☆ Participant, with the participant taking a number of different roles.

Structured or Unstructured

In **structured** observation, the researcher specifies in detail what is to be observed and how the measurements are to be recorded. It is appropriate when the problem is clearly defined and the information needed is specified.

In **unstructured** observation, the researcher monitors all aspects of the phenomenon that seem relevant. It is appropriate when the problem has yet to be formulated precisely and flexibility is needed in observation to identify key components of the problem and to develop hypotheses. The potential for bias is high. Observation findings should be treated as hypotheses to be tested rather than as conclusive findings.

Disguised or Undisguised

In **disguised** observation, respondents are unaware they are being observed and thus behave naturally. Disguise is achieved, for example, by hiding, or using hidden equipment or people disguised as shoppers.

In **undisguised** observation, respondents are aware they are being observed. There is a danger of the Hawthorne effect – people behave differently when being observed.

Natural or Contrived

Natural observation involves observing behaviour as it takes place in the environment, for example, eating hamburgers in a fast food outlet.

In **contrived** observation, the respondents' behaviour is observed in an artificial environment, for example, a food tasting session.

Personal

In personal observation, a researcher observes actual behaviour as it occurs. The observer may or may not normally attempt to control or manipulate the phenomenon being observed. The observer merely records what takes place.

Mechanical

Mechanical devices (video, closed circuit television) record what is being observed. These devices may or may not require the respondent's direct participation. They are used for continuously recording on-going behaviour.

Non-participant

The observer does not normally question or communicate with the people being observed. He or she does not participate.

Participant

In participant observation, the researcher becomes, or is, part of the group that is being investigated. Participant observation has its roots in ethnographic studies (study of man and races) where researchers would live in tribal villages, attempting to understand the customs and practices of that culture. It has a very extensive literature, particularly in sociology (development, nature and laws of human society) and anthropology (physiological and psychological study of man). Organisations can be viewed as 'tribes' with their own customs and practices.

The role of the participant observer is not simple. There are different ways of classifying the role:

☆ Researcher as employee.
☆ Researcher as an explicit role.
☆ Interrupted involvement.
☆ Observation alone.

Researcher as Employee

The researcher works within the organisation alongside other employees, effectively as one of them. The role of the researcher may or may not be explicit and this will have implications for the extent to which he or she will be able to move around and gather information and perspectives from other sources. This role is appropriate when the researcher needs to become totally immersed and experience the work or situation at first hand.

There are a number of dilemmas. Do you tell management and the unions? Friendships may compromise the research. What are the ethics of the process? Can anonymity be maintained? Skill and competence to undertake the work may be required. The research may be over a long period of time.

Researcher as an Explicit Role

The researcher is present every day over a period of time, but entry is negotiated in advance with management and preferably with employees as well. The individual is quite clearly in the role of a researcher who can move around, observe, interview and participate in the work as appropriate. This type of role is the most favoured, as it provides many of the insights that the complete observer would gain, whilst offering much greater flexibility without the ethical problems that deception entails.

Interrupted Involvement

The researcher is present sporadically over a period of time, for example, moving in and out of the organisation to deal with other work or to conduct interviews with, or observations of, different people across a number of different organisations. It rarely involves much participation in the work.

Observation Alone

The observer role is often disliked by employees since it appears to be 'eavesdropping'. The inevitable detachment prevents the degree of trust and friendship forming between the researcher and respondent, which is an important component in other methods.

Choice of Roles

The role adopted depends on the following:

☆ **Purpose of the research**: Does the research require continued longitudinal involvement (long period of time), or will in-depth interviews, for example, conducted over time give the type of insights required?

☆ **Cost of the research**: To what extent can the researcher afford to be committed for extended periods of time? Are there additional costs such as training?

☆ **The extent to which access can be gained**: Gaining access where the role of the researcher is either explicit or covert can be difficult, and may take time.

☆ **The extent to which the researcher would be comfortable in the role**: If the researcher intends to keep his identity concealed, will he or she also feel able to develop the type of trusting relationships that are important? What are the ethical issues?

☆ **The amount of time the researcher has at his disposal**: Some methods involve a considerable amount of time. If time is a problem alternate approaches will have to be sought.

Case-Studies

The term case-study usually refers to a fairly intensive examination of a single unit such as a person, a small group of people, or a single company. Case-studies involve measuring what is there and how it got there. In this sense, it is historical.

It can enable the researcher to explore, unravel and understand problems, issues and relationships. It cannot, however, allow the researcher to generalise, that is, to argue that from one case-study the results, findings or theory developed apply to other similar case-studies. The case looked at may be unique and, therefore not representative of other instances. It is, of course, possible to look at several case-studies to represent certain features of management that we are interested in studying. The case-study approach is often done to make practical improvements. Contributions to general knowledge are incidental.

The case-study method has four steps:

1. Determine the present situation.
2. Gather background information about the past and key variables.
3. Test hypotheses. The background information collected will have been analysed for possible hypotheses. In this step, specific evidence about each hypothesis can be gathered. This step aims to eliminate possibilities which conflict with the evidence collected and to gain confidence for the important hypotheses. The culmination of this step might be the development of an experimental design to test out more rigorously the hypotheses developed, or it might be to take action to remedy the problem.
4. Take remedial action. The aim is to check that the hypotheses tested actually work out in practice. Some action, correction or improvement is made and a re-check carried out on the situation to see what effect the change has brought about.

The case-study enables rich information to be gathered from which potentially useful hypotheses can be generated. It can be a time-consuming process. It is also inefficient in researching situations which are already well structured and where the important variables have been identified. They lack utility when attempting to reach rigorous conclusions or determining precise relationships between variables.

Diaries

A diary is a way of gathering information about the way individuals spend their time on professional activities. They are not about records of engagements or personal journals of thought! Diaries can record either quantitative or qualitative data, and in management research can provide information about work patterns and activities.

Advantages

☆ Useful for collecting information from employees.
☆ Different writers compared and contrasted simultaneously.
☆ Allows the researcher freedom to move from one organisation to another.
☆ Researcher not personally involved.
☆ Diaries can be used as a preliminary or basis for intensive interviewing.
☆ Used as an alternative to direct observation or where resources are limited.

Disadvantages

- ☆ Subjects need to be clear about what they are being asked to do, why and what you plan to do with the data.
- ☆ Diarists need to be of a certain educational level.
- ☆ Some structure is necessary to give the diarist focus, for example, a list of headings.
- ☆ Encouragement and reassurance are needed as completing a diary is time-consuming and can be irritating after a while.
- ☆ Progress needs checking from time-to-time.
- ☆ Confidentiality is required as content may be critical.
- ☆ Analyses problems, so you need to consider how responses will be coded before the subjects start filling in diaries.

Chapter 19

Training of Leaders

Meaning of Training

The term 'training' is used to those activities aimed at improving the ability of a person to do his job including acquainting (to know) information, developing abilities, attitudes that will result in greater professional competency. The potential leaders who are selected by various methods, **lack** some of the essential **traits** of leadership, the qualities can be developed by training objectives as follows:

Objectives of Training of Leaders

1. The objective of training is to develop the essentials of **good leadership** in the selected leaders

2. To give them a perfect understanding of the people, to enable them to understand **group behavior**

3. Develop competence in group processes *i.e.* teaching them the methods of **identifying** problems develop cooperative thinking, exchange and analysis of ideas

4. To acquire **technical skills** necessary to carry out a **job**, how to identify problems and plan appropriate procedures. To obtain attitudes, knowledge and skills of dealing with people. To develop in them latest capacities of leadership

Methods of Training of Professional Leaders

1. **Background courses in college or an institution**: Giving training on general college education in a college or an institution in **psychology** or **sociology**

2. **Induction training**: apprenticeship experience under the direction of a trained and experienced leader in the field will enable the new professional leader to develop his abilities for successful leadership

3. **In-service training**: This is training is given to the professional leaders for constantly **improving** their **efficiency** by focusing attention upon the problems they have faced in the field and the ways to solve them. In-service training has become increasingly **important** in view of the fast changing **technology** in agriculture in recent times

Methods of Training of Lay Leaders

The different methods of training lay leaders are classified in to **two** types one is **formal** and the other is **informal** as given below:

Formal

1. Lecture
2. Discussion
3. Symposium
4. Workshop
5. Forum
6. Panel
7. Field trip
8. Apprenticeship
9. Training camps
10. Direct assistance from experts
11. Buzz groups
12. Giving responsibility to local leaders
13. Audio-visuals

Informal

1. Observations
2. Reading
3. Talking

Formal Methods of Training of Lay Leaders

1. **Lecture**: This is probably **most common** method. Through this method local leaders under training are given enough **material** for thought, but little opportunity for **self-expression.** The lecture method is effective in certain situations, but usually is supplemented by other methods, depending on the objectives to be attained

2. **Discussion**: Discussion usually occurs in a **face to face** or co-acting situation in which people involved, exchange the useful information by speaking with each other

3. **Workshop**: It is essentially a long-term meeting form one day to several weeks, involving all the delegates (participants) in which problems are discussed by delegates in small private groups. The workshop as the name indicates must produce **something** in the end a report, a publication, a visual or any other material object

4. **Forum**: it is assemble (group of people) for discussion of matters of interest and usually follows the other extension teaching methods. In the forum the audience clear their doubts and raise **questions** for additional information

5. **Panel**: it is **informal conversation** for the benefit of the audience by a small group of speakers, usually from **2** to **8** in number

6. **Symposium**: this is **short series** of **lectures** in which **3** or **4** speakers explain the **different parts** of a particular subject

7. **Field trip**: in this method a group people go to see and gain firsthand knowledge of improved practices in their natural setting

8. **Apprenticeship**: in this the local leaders or the potential leaders see **someone** operating with a view to learn some of the activities and ways of handling the problems in the field of leadership

9. **Training camps**: Training is imparted by organizing **camps** in which several local leaders are involved in the training sessions at the same time

10. **Direct assistance from experts**: this may come in the form of advice from an expert in the field of leadership

11. **Buzz groups**: in this a large group is divided into smaller units for a short period called **buzz session.** It is also called as **huddle system** or **Phillips 66** in which group of **6** to **8** persons get together after receiving instructions to discuss about a specific issue assigned

12. **Giving responsibility** to local leaders: giving everyone a job by which self confidence may be attained by achievement in activities useful to the group is essential for development of leadership

13. **Audio-visuals**: These include role playing, socio-drama, demonstration, movies *etc.*

Informal Methods of Training Lay Leaders

1. **Observation**: Noticing how others have performed through observation
2. **Reading**: Studying printed material often found in the form of leader hand-books, newsletters, circulars, bulletins *etc.*
3. **Talking**: Speaking with other leaders in the same or related fields of interest and also with members to determine consensus (common opinion)

Advantages of using Local Leaders in Extension

1. Local leaders act as **extension teachers** and this helps in increasing the adoption of improved practices
2. **Cost** of extension is reduced as local leaders are **not paid** for their work

3. Local leaders themselves become **better taught,** because of the experience they gain in teaching and influencing others

4. People accept new idea **more readily** form a local person who has practically tried it, while they may **resist** if the ideas were to come from an extension worker

5. The frequent contacts of extension workers with local leaders raises his **prestige** thereby making him more effective in his work

Limitations of using Local Leaders in Extension

1. Person selected as leader may not have the expected **following** among neighbours or may not be willing to **devote required time** to work, or may be a **poor** teacher

2. Considerable time is required to locate and train local leaders

3. Local leader may try to use prestige connected with position of personal advantage

4. The most difficult task of arousing interest on the part of those not interested in extension is too often left to the in experienced local leader

5. Public recognition and publicity given to informal local leaders may sometimes jeopardize (spoil) their position and adversely affect their influence.

Author Index

Subject Index